Economic Evaluation
of Hydropower and Water Conservancy Projects
—Theory and Practice

水利水电建设项目经济评价及案例分析

王晓云 刘武军 王辉 李春玲 曾俊 译

中国水利水电出版社
www.waterpub.com.cn
·北京·

Brief Introduction

Focused on water conservancy and hydropower projects, this book introduces the related economic evaluation, financial evaluation, fund sources and financing plan, uncertainty analysis, cost allocation, as well as calculation methods of financing costs and on-grid price of overseas projects financed from China.

This book is written based on *Construction Projects Economic Evaluation Method and Parameter* (3rd Edition) (National Development and Reform Commission, Ministry of Construction, 2006), *Regulation for Economic Evaluation of Water Conservancy Construction Projects* (SL 72-2013) and *Specification on Economic Evaluation of Hydropower Project* (DL/T 5441-2010). It can be used as a reference book for project designers, decision makers of project investment or beginners of project economic evaluation on the study and practice of the three manual books.

图书在版编目（ＣＩＰ）数据

水利水电建设项目经济评价及案例分析 ＝ Economic Evaluation of Hydropower and Water Conservancy Projects-Theory and Practice：英文 ／ 王晓云等译. －－ 北京：中国水利水电出版社，2022.6
ISBN 978-7-5226-1127-3

Ⅰ.①水… Ⅱ.①王… Ⅲ.①水利水电工程－基本建设项目－项目评价－案例－中国－英文 Ⅳ.①F426.9

中国版本图书馆CIP数据核字（2022）第219180号

书　　名	Economic Evaluation of Hydropower and Water Conservancy Projects—Theory and Practice 水利水电建设项目经济评价及案例分析 SHUILI SHUIDIAN JIANSHE XIANGMU JINGJI PINGJIA JI ANLI FENXI
译　　者	王晓云　刘武军　王辉　李春玲　曾俊　译
出版发行	中国水利水电出版社 （北京市海淀区玉渊潭南路1号D座　100038） 网址：www.waterpub.com.cn E - mail：sales@mwr.gov.cn 电话：（010）68545888（营销中心）
经　　售	北京科水图书销售有限公司 电话：（010）68545874、63202643 全国各地新华书店和相关出版物销售网点
排　　版	中国水利水电出版社微机排版中心
印　　刷	天津嘉恒印务有限公司
规　　格	184mm×260mm　16开本　19印张　601千字
版　　次	2022年6月第1版　2022年6月第1次印刷
印　　数	001—500册
定　　价	120.00元

凡购买我社图书，如有缺页、倒页、脱页的，本社营销中心负责调换

版权所有·侵权必究

Foreword

With a vast territory, China is rich in water resources. However, the spatial and temporal distribution of water resources is uneven due to the integrated impacts of climate, topography, and other factors. Specifically, precipitation decreases from south to north. The interannual variation of yearly precipitation is significant. Frequent floods and droughts occurred and observed. Since 1949, many conservation and hydropower engineering project, eg., water storage, diversion, and transfer projects have been constructed, which greatly contribut to the economic and social development. Currently, higher requirements have been put forward for water conservancy projects to support a sustainable, rapid socio-economic development, as well as the strategic needs of building a moderately prosperous society in an all-round way. In 2014, the State Council arranged 172 major water conservancy projects for water saving and water supply, providing new opportunities for water conservancy construction.

China is rich in hydraulic China is the world's largest producer of hydroelectricity. In the mainland, the potential annual hydroelectric power is 6.08 trillion kW·h, the average power is 694 million kW; technical developable installed capacity is 542 million kW, annual power generation is 2.47 trillion kW·h; economically developable installed capacity is 402 million kW, annual power generation is 1.75 trillion kW·h. Presently in China, the installed capacity of hydropower is over 230 million kW, ranking first in the world. Hydropower, as a renewable clean energy, has the characteristics of mature technology, low cost, flexible operation and so on. Most countries in the world give priorities to hydropower development in energy construction. Meanwhile, China leads the world in renewable-energy production mainly from hydropower. Hydropower has made significant contributions in many aspects, For example, it helps to adjust the energy and power structure, reduce greenhouse gas emission, mitigate the climate and environment, and socio-economic development. Therefore, hydropower is listed in the key areas of energy development and renewable energy development in China.

The global potential annual hydroelectric power is about 43.6 trillion kW·h, 15.8 trillion kW·h for technology development, and 9.3 trillion kW·h for economic exploitable annual electricity generation. Until recently, about 25% of the global hydropower has been developed. Specifically, the developed ratios in Europe, North America, South America, Asia, and Africa are 47%, 38%, 24%, 22% and 8%, respectively. In the future, the key areas for hydropower constructions and development are Asia, Africa, and South America. With the implementation of the "One Belt and One Road" and the hydropower going-global strategy, the construction of water conservancy and hydropower projects will be set off in related areas.

Economic evaluation plays an important role in early decision-making of water conservancy and hydropower projects. Economic evaluation is to investigate, predict, study, calculate and demonstrate the input and output factors with modern economic analysis methods, during the calculation period of the proposed project. Finally, it provides a comprehensive analysis and evaluation on the economic rationality and financial feasibility of the project. And the purpose is to find the optimal economic benefits with possible funds. With the rapid development of water conservancy and hydropower projects and gradually in China, accelerated: foreign investment in China, project decision makers and consultants are encountering a primary problem. That is, how to achieve a comprehensive understanding, master, and judgement of the financial situation of the project, and make a scientific investment decision to reach an optimal investment efficiency.

In order to promote and standardize water conservancy and hydropower construction projects, two important manuals were issued based on *Construction Projects Economic Evaluation Method and Parameter (3rd Edition)*. In details, the Ministry of Water Resources of the People's Republic of China issued *Regulation for Economic Evaluation of Water Conservancy Construction Projects (SL 72 -2013)*, and National Energy Administration issued the *Specification on Economic Evaluation of Hydropower Project (DL/T 5441 - 2010)*. Currently, higher requirements have been put forward for project economic evaluation and financial evaluation. This is necessary and urgent with the speeding-up investment and financial requirements in water conservancy and hydropower projects at home and abroad, the implementation of PPP model, as well as requirements of project owners, creditors, etc. This book is compiled based on personal and professional experiences in project design, review and evaluation of water conservancy and hydropower. Also, it follows the specification of *Construction Projects Economic Evaluation Method and Parameter* (3rd Edition), *Specification on Economic Evaluation of Hydropower Project* (DL/T 5441-2010) and *Regulation for Economic Evaluation of Water Conservancy Construction Projects* (SL 72-2013).

This book consists of eleven chapters.

Chapter 1 introduces the purpose and target of economic evaluation of water conservancy and hydropower projects, as well as the main terms used in this book.

Chapter 2 is on time value of capital and equations for cash flow equivalents. Specifically, this chapter mainly introduces the calculation of time value of capital, and the calculation method of cash flow equivalence.

Chapter 3 is on national economic evaluation. Specifically, this chapter mainly introduces the estimation principles of economic benefit and cost identification, the estimation methods of national economic costs, the estimation methods of various engineering benefits, and the parameter values.

Chapter 4 is on financial evaluation of water conservancy andhydropower projects. Specifically, this chapter mainly introduces the basic principle and the price system of financial evaluation, the estimation methods of total investment and total cost, the parameter value, the financial form calculation, and the financial evaluation index, etc.

Chapter 5 is on fund sources and financing plan. Specifically, this chapter mainly introduces the fund sources and the formulation of financing schemes. Besides, this chapter illustrates the calculation of loan capacity and analysis of capital structure of water conservancy and hydropower projects through cases.

Chapter 6 is on uncertainty analysis of water conservancy and hydropower projects. Specifically, sensitivity analysis, break-even analysis, and risk analysis are mainly introduced in this chapter. Besides, the risk of water conservancy and hydropower projects is included.

Chapter 7 is on economic comparison of project schemes. Specifically, this chapter mainly introduces the principles and methods of economic comparison of different schemes of water conservancy and hydropower projects.

Chapter 8 is on cost allocation. Specifically, this chapter mainly introduces the investment allocation principle of multi-purpose key projects, and the division of shared projects and special projects. Also, it introduces the investment allocation methods through the cases.

Chapter 9 provides case study of economic evaluation of hydropower projects in China. Specifically, this chapter mainly introduces the basic data, national economic evaluation and calculation, on-grid price calculation, financial evaluation calculations using Excel-internal functions, and financial evaluation of hydropower projects.

Chapter 10 case study of economic evaluation of water conservancy projects. Specifically, this chapter introduces the basic data, national economic evaluation and calculation, investment sharing and total cost allocation, cost analysis, market analysis and user affordability analysis, loan capacity calculation, water conservancy project financial evaluation form calculation, and financial evaluation, etc.

Chapter 11 provides foreign case of financial evaluation of water conservancy and hydropower projects financed from China. For example, it introduces calculation methods and the structure of electricity price in Pakistan. Besides, the Excel-internal functions for economic evaluation and calculation are also introduced in this book.

This book can be used as a reference book for the study of *Construction Projects Economic Evaluation Method and Parameter (3rd Edition)*, *Specification on Economic Evaluation of Hydropower Project (DL/T 5441-2010)* and *Regulation for Economic Evaluation of Water Conservancy Construction Projects (SL 72-2013)*.

Due to the limited time and the authors' level, there might still exist some errors. Criticisms and suggestions are welcome and appreciated.

The authors
July 2021

Contents

Foreword

1 Purpose and Target of Economic Evaluation ·········· 1
  1.1 Purpose and Target of Economic Evaluation ·········· 1
  1.2 The Main Differences Between National Economic Evaluation and Financial Evaluation ·········· 2
  1.3 Main Notations ·········· 3

2 Time Value of Capital and Equations for Cash Flow Equivalents ·········· 10
  2.1 Time Value of Capital and Cash Flow Equivalents ·········· 10
  2.2 Basis Point and Cash Flow Diagram ·········· 13
  2.3 Basic Formula of Dynamic Capital ·········· 14

3 National Economic Evaluation ·········· 25
  3.1 Requirements and Estimation Principles of Economic Benefit and Cost Identification ·········· 25
  3.2 Cost Calculation ·········· 26
  3.3 Benefit Calculation ·········· 28
  3.4 National Economic Evaluation Indicators ·········· 68
  3.5 National Economic Evaluation ·········· 69

4 Financial Evaluation ·········· 72
  4.1 Basic Principles, Price System and Evaluation Contents of Financial Evaluation ·········· 72
  4.2 Total Investment Estimation ·········· 75
  4.3 Total Cost Estimation ·········· 83
  4.4 Water Supply Price and On-grid Price ·········· 100
  4.5 Financial Evaluation ·········· 107

5 Fund Sources and Financing Plan ·········· 135
  5.1 Fund Sources ·········· 135
  5.2 Financing Plan ·········· 137

6 Uncertainty Analysis ·········· 143
  6.1 Sensitivity Analysis ·········· 143

  6.2 Break-even Analysis ·· 148
  6.3 Risk Analysis ·· 151
7 Economic Comparison of Project Schemes ·· 154
  7.1 Scheme Comparability ·· 154
  7.2 Relationship among Schemes ··· 155
  7.3 Comparison Methods of Mutually Exclusive Schemes ····························· 156
8 Cost Allocation ··· 162
  8.1 Investment Classification of Multi-purpose Water Conservancy Construction Projects ··· 162
  8.2 Principle of Cost Allocation ·· 163
  8.3 Investment Allocation in Special Projects and Projects of Common Purpose ··· 163
  8.4 Cost Allocation Method of Works for Common Purpose ·························· 165
  8.5 Rationality Analysis of Cost Allocation ·· 172
9 Case Study of Economic Evaluation of Hydropower Projects ······················· 174
  9.1 Project Overview ·· 174
  9.2 Purpose and Basis of Economic Evaluation ·· 175
  9.3 National Economic Evaluation ·· 175
  9.4 Financial Evaluation ·· 181
  9.5 Conclusions and Suggestions ··· 203
10 Case Study of Economic Evaluation of Water Conservancy Projects ············ 205
  10.1 Project Overview ·· 205
  10.2 Evaluation Purpose and Basis ·· 206
  10.3 Fixed Assets Investment and Total Cost ··· 206
  10.4 National Economic Evaluation ·· 210
  10.5 Cost Sharing and Cost Estimation ··· 218
  10.6 Market Situation and User Affordability Analysis ································· 223
  10.7 Measurement of Loaning Ability ·· 227
  10.8 Financing Schemes ·· 233
  10.9 Financial Evaluation ··· 235
  10.10 Conclusions ·· 259
11 Foreign Case of Financial Evaluation of Water Conservancy and Hydropower Projects ··· 260
  11.1 Project Overview ·· 260
  11.2 Purpose and Basis of Financial Evaluation ·· 261
  11.3 Tariff Structure of Pakistan ··· 262

11.4	Total Investment and Fund Raising of the Project	263
11.5	Total Cost Estimation	265
11.6	Calculation of Generation Income	265
11.7	Taxes	266
11.8	Capital Return Requirements	266
11.9	Financial Evaluation	267
11.10	Sensitivity Analysis	289
11.11	Conclusions	290

References 291

1

Purpose and Target of Economic Evaluation

1.1 Purpose and Target of Economic Evaluation

The services of water conservancy and hydropower projects include flood control, waterlogging control, irrigation, water supply, hydropower generation, shipping, soil and water conservation, ecological restoration, interbrain water transfer, etc. There are both single-purpose projects and multi-purpose projects (i.e., comprehensive utilization projects).

Economic evaluation of water conservancy and hydropower construction projects is to comprehensively evaluate the economic rationality and financial feasibility of the projects. The purpose is to recommend an optimal scheme through scheme comparison by use of modern economic analysis methods. Specifically, it performs investigation, prediction, study, calculation and demonstration on the various factors (i.e., the input and output) considered in the calculation period of the proposed projects.

Specifically, economic evaluation of projects includes national economic evaluation and financial evaluation.

As for national economic evaluation of the projects, the main purpose is to provide the government sectors with scientific basis on project approval, project appraisal, as well as scheme comparison and selection. Its functions include: ① evaluate project economic rationality and its net contribution to social economy; ② provide basis for government rational resources allocation; ③ provide basis for project examination and approval by government; ④ provide basis for the formulation of financial scheme for projects such as market-oriented infrastructure; ⑤ help to reach a balance between the enterprises and the whole society on interests; ⑥ scheme comparison and selection.

As for financial evaluation, the main purpose is to provide comprehensive financial information of projects to project owners and project creditor organization so as to make investment decisions and judge the project solvency and debt risk. In details, financial viability, solvency and profitability should be provided. Specific roles of financial evaluation include: ①important part in project decision analysis and project evaluation; ②important decision-made basis for project initiation; ③important in scheme comparison and selection; ④important in the evaluation of project financial sustainability.

There are three main targets for the economic evaluation of water conservancy and hydropower construction projects, including: ①analyze and predict various expenses, economic and financial benefits and their development trends of the construction projects; ②carry out the national economic evaluation and financial evaluation, evaluate the economic internal rate of return (EIRR), economic net present value (ENPV) and economic benefit cost ratio (EBCR) of the projects, as well as the financial internal rate of return (FIRR), financial net present value (FNPV), investment payoff period and finance; ③calculate the unit function index of the project.

1.2 The Main Differences Between National Economic Evaluation and Financial Evaluation

For water conservancy and hydropower construction projects, national economic evaluation analyzes the cost of investment activities and the economic effects produced by occupation of economic resources. Also, it evaluates the resource allocation efficiency from the perspective of the whole society. Comparatively, financial evaluation is to judge the financial feasibility of projects. It calculates financial benefits and costs within project scope based on current national fiscal and taxation system, as well as price system. The differences between national and financial economic evaluations are as follows:

1. Different in evaluation levels

National economic evaluation is from the perspective of the overall interests of the national (social) economy, to investigate the contribution of the project to the national economy, analyze the economic efficiency, effect and impact of the project on the society, and evaluate the rationality of the project in the macro economy.

Financial evaluation is under the premise of the current national fiscal and taxation system and price system, from the perspective of the project, to calculate the financial expenditure and income within project scope, analyze financial viability, solvency and profitability of projects, and evaluate financial feasibility of projects.

2. Different in scopes of calculation for cost and benefit

National economic evaluation calculates project costs and benefits from the perspective of the whole society. Therefore, all kinds of subsidies transferred within the national economy do not belong to the benefits, similarly, various taxes not to the cost. Comparatively, financial evaluation calculates financial expenditure and income from the perspective of project finance. Hence, all kinds of taxes paid shall be taken as the financial expenditure, and the subsidies as the income.

Moreover, the national economic evaluation evaluates the external effects of projects. Therefore, it analyzes and calculates indirect costs and benefits of projects. Whereas, the financial evaluation only calculates direct expenditure and income.

3. Different in prices of inputs and outputs

Shadow price is adopted in national economic evaluation, whereas financial price is adopted in financial evaluation. Shadow price is determined according to some principles, and more reasonable than financial price. It can better reflect real value of products, market supply and demand, scarcity of resources, and realize optimized resource allocation. Financial price is the forecast price based on current price system. It has three price forms such as national pricing, national guidance price and market price. In the case of coexistence of various prices, the financial price shall be the price that most likely to occur.

4. Different in key parameters

In national economic evaluation, shadow exchange rate and social discount rate which are measured uniformly by the state, are used. However, in financial economic evaluation, national foreign exchange rate quotation and the industry financial benchmark yield are used.

1.3 Main Notations

1. Calculation period, construction period and operation period

Calculation period is the period set for economic and financial analysis in national economic evaluation and financial evaluation. It includes the construction period and operation period. Construction period is the time required from the formal investment of project funds to the completion and operation of the project. Operation period is the period from the completion of the project to reach the design life.

In some large hydropower projects, the power station units are put into operation gradually. The period from the first unit to the completion of the project is called the initial operation period. That is, the initial operation period is included in the construction period. Water conservancy projects can only generate benefits after project completion. When the project has just been completed, the water demand is less than the designed water supply. This period is called the initial operating stage.

2. Total investment, fixed assets investment, construction investment, fixed assets value, and original value of fixed assets

The total investment of water conservancy project is different from that of hydropower project. The total investment of hydropower project is the sum of construction investment, interest during construction and working capital. Comparatively, the total investment is the sum of fixed assets investment and interest during construction. The estimated investment of hydropower projects is called construction investment. It includes the investment of hub project, compensation for land acquisition and resettlement, independent cost and reserve cost. Comparatively, the estimated investment in water conservancy projects is called fixed asset investment. It includes project investment, resettlement and environment, water and soil conservation investment, and reserve fund.

According to the principle of capital preservation, when project is completed and put into operation, the construction investment and the interest during the construction period form three parts: fixed assets, intangible assets and other assets, that is, the original value of fixed assets, intangible assets and other assets. For water conservancy construction projects, there are few intangible assets and other assets. The cost of land acquisition for immigrants is embodied in the cost of buildings. The investment of fixed assets and the interest during the construction period are the original value of fixed assets, which can also be called the value of fixed assets or the original value of fixed assets.

3. Total cost, fixed cost, variable cost, depreciation, amortization

The total cost is all expenses incurred to produce products or services during the operation period. It is the sum of the operation cost, depreciation expense, amortization expense and financial expense.

Fixed cost is also known as constant cost. In total cost, it does not change with the increase or decrease of product output.

On the contrary, the variable cost is the part of total cost that changes with the increase or decrease of product output.

The depreciation cost is also known as the depreciation cost of fixed assets. The fixed assets will be worn out during project operation. Simultaneously, the loss of value is usually compensated by the way of drawing depreciation. According to financial system provisions, the fixed assets shall be depreciated on monthly basis, and shall be included in the cost or current profit and loss of relevant assets. In financial analysis, when the total cost is estimated by the factor of production method, the depreciation of fixed assets can be directly listed in the total cost. Methods of the depreciation of fixed assets can be determined by the enterprises according to tax law. Generally, the straight – line method is adopted, including age limit method of average and the workload method. In China, the rapid depreciation method is also allowed by tax law for certain machinery and equipment. Specific methods are the double declining balance method and the total number of years method. For water conservancy construction projects, the age limit method of average, as a straight – line method, is generally adopted. The annual depreciation cost is the product of the value of fixed assets and the annual depreciation rate.

Amortization expenses is the amortization expenses of intangible assets and other assets. According to relevant regulations, intangible assets shall be amortized and included in the cost averagely from the date of use to effective service life. If the legal term of validity or the beneficial period is stipulated by laws and contracts, the amortization period shall follow its provisions. Otherwise, the amortization period shall meet the requirements of tax law. Generally, the straight-line average method is adopted in the amortization of intangible assets, and the residual value is excluded. All this is the same in the amortization of other assets. That is, the straight-line average method is also adopted in the amortization of other assets, and the residual value is excluded. Also, the amortization period should meet the requirements of tax law. The amortization fee of intangible assets and other assets is equal to the product of intangible assets value, other assets value and the rate of amortization fee. In water conservancy projects or hydropower projects, there are few intangible assets and other assets. Therefore, the intangible assets and other assets are not taken into account.

4. Investment in operation maintenance (renovation investment)

The investment in operation maintenance (renovation investment) is the needed investment to maintain normal operation, specifically it is a certain amount of fixed assets. For water conservancy and hydropower construction projects, this investment is mainly used for replacement of electromechanical, metal structure and other equipment after effective service life.

5. Time value, static analysis and dynamic analysis of capital

The increasing value of money over time is termed as the time value of money.

Static analysis is the analysis method of capital without considering the time value of money. The indicators of static analysis include static investment payback period, total investment return rate, capital net profit rate and others.

Dynamic analysis is the analysis method of capital considering the time value of money. The indicators of dynamic analysis include dynamic investment payback period, internal rate of return (IRR), net present value (NPV), benefit-cost ratio (R_{BC}), etc.

6. Interest rate, interest, nominal interest rate, and effective annual interest rate

Interest rate is the interest generated by unit capital during an interest period.

Interest is the price paid for the occupation of funds (loan interest) or the compensation received for abandoning the use of funds (deposit interest).

Nominal interest rate is the interest rate published by the central bank or other institutions providing funds for lending.

The effective annual interest rate is the interest rate converted to year according to interest period such as quarter, month and day.

7. National economic evaluation

National economic evaluation is performed from the perspective of the whole society. The cost of investment activities is analyzed. The economic effects of project occupying economic resources, and resource allocation efficiency of project investment are evaluated. In national economic evaluation, shadow price, shadow exchange rate, shadow wage and social discount rate are used.

8. Financial evaluation

Financial evaluation is performed from project perspective under the premise of the current national financial and tax system and price system. Specifically, the financial benefits and costs are calculated, and financial statements are prepared. Evaluation indicators are calculated, and the viability, solvency and profitability of projects are examined. All these are provided to judge financial appraisal of projects, and provide basis for decision making.

9. Social discount rate and financial benchmark yield

Social discount rate is a general parameter of national economic evaluation of construction projects, and it is the main basis for judging economic feasibility of construction projects. Social discount rate represents the minimum rate of return required by social investment. According to *Construction Projects Economic Evaluation Method and Parameter (3rd Edition)*, the recommended social discount rate is 8%.

In national economic evaluation, the social discount rate should be set as 8%, the state - stipulated value, according to *Regulation for Economic Evaluation of Water Conservancy Construction Projects (SL 72 - 2013)*. Water conservancy projects, e.g., flood control project, drainage project, and ecological management project, have unique functions of social public benefits including social benefits, environmental benefits, and other benefits. However, these benefits are difficult to reflect in currency. Therefore, the social discount rate at 6% can be acternatively adepted in economic evaluation.

Financial benchmark yield is the benchmark rate of return for calculating the financial net present value of financial costs and benefits in project financial evaluation with the discount method. It is the benchmark value for measuring the internal rate of return of the financial rights of the project and the main criterion for the comparison and selection of the financial feasibility and scheme of the project. In essence, it reflects the investors' judgment on the time value of money and the estimation of the risk degree of the project. According to *Construction Projects Economic Evaluation Method and Parameter (3rd Edition)*, the pre - tax financial benchmark yield of reservoir power generation project before financing is 7%. The post - tax financial benchmark yield of project capital is 10%. The pre - tax financial benchmark yield of water diversion and water supply project before financing is 4%. The post - tax financial benchmark yield of project capital is 6%.

10. Economic internal rate of return (EIRR), economic net present value (ENPV) and economic benefit - cost ratio (EBCR)

Economic internal rate of return (EIRR) is the discount rate when the accumulated net benefit present value of each year in the calculation period of the project is equal to zero.

Economic net present value (ENPV) is the value of all future net benefits discounted to the present value at the beginning of the calculation period, with the social discount rate.

The economic benefit-cost ratio (EBCR or R_{BC}) is the ratio of benefit present value to cost present value of projects during calculation period.

11. Interest coverage ratio, debt service coverage ratio (DSCR) and asset liability ratio

Interest coverage ratio is the ratio of earnings before interest and taxes (EBIT) to interest payable during the loan repayment period. It reflects the guarantee degree of project debt interest repayment from the perspective of the adequacy of interest payment capital source. Some creditors require an interest coverage ratio of no less than 2.

Debt service coverage ratio (DSCR) is the ratio of the capital used to calculate the principal and interest repayment to the total principal and interest payable during the loan repayment period. It reflects the guarantee degree of the available capital to repay the principal and interest of the loan. Some creditors require a debt service coverage ratio of no less than 1.3.

Asset liability ratio is the ratio of total liabilities to total assets at the end of each period.

12. Financial internal rate of return (FIRR), financial net present value (FNPV), and payback period

Financial internal rate of return (FIRR) is the discount rate when the present value of net cash flow in the calculation period of the project is equal to zero.

Financial net present value (FNPV) is the sum of the present value of the net cash flow during project calculation period of the project. It is calculated based on the set discount rate (generally the benchmark yield rate).

Investment payback period is the time required to recover the project investment based on the net income of the project, generally in years. It should be calculated from the year when project construction starts. If it is calculated from the year when project operation starts, this should be noted. The investment payback period can be divided into static and dynamic investment payback periods. The static investment payback period is the time required to compensate the project investment with the net income of each year without considering the time value of money. Comparatively, the dynamic payback period is the time required to compensate the present value of the project investment with the present value of the net income of each year considering the time value of money. Specifically, the investment payback period is the static investment payback period in *Regulation for Economic Evaluation of Water Conservancy Construction Projects* (SL 72 -2013) and *Specification on Economic Evaluation of Hydropower Project* (DL/T 5441 -2010).

13. Total return on investment, net profit margin of project capital

The total return on investment is the profit level of total investment. It is the annual EBIT of the normal year after the project reaches the design capacity, or the ratio of the annual average EBIT during operation period to project total investment.

The net profit margin of project capital refers to the profit level of project capital. It is the ratio of, the annual net profit or the annual average net profit during operation period, to the project capital.

14. Capital fund, debt fund and financing plan

Capital fund is the amount of capital contribution subscribed by the investor in the total investment of the construction project. It is non debt capital of the construction project, so the project legal person does not bear any interest and debt. The investor can enjoy the owner's equity in accordance with the proportion of capital contribution. Also, the investor can make a transfer of the capital contribution, but generally it is not allowed to withdraw in any way. The capital sources include government capital at all levels, government subsidies, government investment, capital invested by project enterprise legal person, personal capital and other funds.

Debt fund (or debt capital) is the capital of project investment which are obtained from financial institutions, securities markets and other capital markets in the form of liabilities. Debt capital has time limit. Principal and interest should be repaid at regular intervals. Moreover, the capital cost is generally lower than equity capital, ensuring that the investors' control over the enterprise will not be dispersed.

Financing plan designs the proportions of capital and debt in project investment. For power generation projects, the minimum proportion of capital is 20%. Whereas, for urban water supply (or water transfer projects), the minimum proportion of capital shall not be less than 35%.

15. Measurement of maximum loan capacity

The measurement of the maximum loan capacity is to reasonably predict the financial income and expenses of projects according to market demand.

It measures the maximum loan amount and required capital amount that can maintain project operation under current financial tax system with qualified bank credit.

16. Sensitivity analysis and break-even analysis

Sensitivity analysis is a kind of uncertainty analysis method. At first, identify sensitive factors from many uncertain factors, which have important influence on economic benefit indicators of the investment project. Then, analyze and calculate the influence and sensitivity of the sensitive factors on economic benefit indicators. Next, judge risk tolerance of the project.

Break-even analysis is also called volume-cost-profit analysis. The analysis is conducted according to the product output (sales volume), fixed cost, variable cost, product price and tax in the normal year. The break-even point of the project is calculated. Also, the change and balance relationship among the project output, cost and profit is analyzed.

17. Risk and risk analysis

Risk is the possibility of deviation from expected financial and economic benefits because of uncertain factors after the implementation of water conservancy and hydropower construction projects.

Risk analysis includes: Firstly, identify risk factors, using qualitative or quantitative analysis methods. Secondly, estimate the occuring possibility of identified risk factors and their impacts on the project. Thirdly, investigate reveals the key risk factors affecting the project. Finally put forward effective countermeasures.

18. Cost allocation

Cost allocation is the allocation of project investment and cost of comprehensive utilization project to each beneficiary area or each function according to the proportion of income or other allocation methods.

2

Time Value of Capital and Equations for Cash Flow Equivalents

2.1 Time Value of Capital and Cash Flow Equivalents

2.1.1 Time Value of Capital

Time value of capital is the increasing value of capital over time. It is also called time value of money, as money is the main form of capital. Capital is put into production as a factor of production, combined with labor, to produce necessary labor value and surplus labor value (profit). The former is the value of capital itself, and the latter is the value increment of capital, which is the so-called time value of capital.

The value of capital increases over time. But it does not mean money itself can increase value. However, money is the material manifestation of labor value. Only combined with labor in production and circulation, can money add value.

2.1.2 Interest and Interest Rates

The time value of capital is closely related to interest and expressed by an interest rate. Interest and interest rate are two basic indicators of the time value of capital. Interest is the price paid for the occupation of funds (loan interest) or the compensation received for abandoning the use of funds (deposit interest). Interest rate is the interest generated by unit fund during unit interest period.

$$interest\ rate = \frac{interest\ increased\ per\ unit\ interest\ period}{principal}$$

Interest is not only closely related to principal, interest rate and duration, but also related to the length of the interest period. It can be calculated on a daily, monthly, quarterly or annual basis. Interest rate is expressed in different ways given different interest periods or habits. It is usually expressed as the interest rate in a commonly used interest period.

2.1 Time Value of Capital and Cash Flow Equivalents

There are two methods to calculate capital interest, including: simple interest method, and compound interest method.

1. Simple interest method

In simple interest method, the base of interest calculation is only the principal in each interest period, and the interest calculated in the previous interest period is not considered. In other words, the interest is not added to the principal for calculating interest. Consequently, the interest of a certain amount of principal in every interest period is the same. In this method, the sum of the principal and interest can be computed using the following formula:

$$F = P(1 + i \cdot n)$$

Where: F—the sum of the principal and interest;

P—initial principal amount;

i—interest rate;

n—number of interest periods.

Currently in banks of China, the simple interest method is generally adopted in calculating deposit interest.

2. Compound interest method

In each interest period, the computation base is the principal plus the interest computed in the previous period. In other words, the amount of interest for a period is added to the amount of principal to compute the interest for next period. Therefore, compound interest method is widely known as "snowball interest". In this method, the final amount can be computed using the following formula:

$$F = P(1 + i)^n$$

F, P, i and n have the same meaning as nominated above.

The bank loans in China are calculated by compound interest method. Generally, compound interest method is adopted to calculate the interest in the economic analysis of projects.

2.1.3 Nominal and Effective Interest Rates

The interest rate stated by each financial institution for one-year loan and above is on annual basis. However, in actual economic activities, the interest may be compounded monthly, quarterly, semi-annually or annually. With these different periods, the accumulated interest at the end of the year will be different for the same annual loan interest rate. Therefore, there exist nominal interest rate and effective interest rate.

Nominal interest rate is the interest rate stated by the central bank or other institutions providing capital loans. Comparatively, effective interest rate is the annual interest rate converted from the quarterly, monthly, daily and other interest periods.

Assume that:

P—principal at the beginning of the year;
F—principal plus interest at the end of the year;
L—interest generated in the year;
r—nominal interest rate;
i—effective interest rate;
m—number of interest periods in a year.

Then, the interest rate of unit interest period is r/m, and the principal plus interest at the end of year is as follows:

$$F = P\left(1 + \frac{r}{m}\right)^m$$

The interest generated within one year is:

$$L = F - P = P\left[\left(1 + \frac{r}{m}\right)^m - 1\right]$$

The actual effective interest rate is:

$$i = \left(1 + \frac{r}{m}\right)^m - 1$$

[**Example 2.1**] In order to expand the scale of production, a water conservancy unit borrows money from financial institutions. The annual interest rate is 6%. If the interest is monthly compounded, what is the effective interest rate for repayment? If the interest is compounded semi-annually, what is the effective interest rate for repayment?

Solution: Given $r = 6\%$, and the interest is compounded monthly: Then $n = 12$, and the effective annual interest rate is:

$$i = \left(1 + \frac{r}{n}\right)^n - 1 = \left(1 + \frac{0.06}{12}\right)^{12} - 1 = 6.17\%$$

If the interest is compounded semi-annually, then $r = 6\%$, $n = 2$, and the effective annual interest rate is:

$$i = \left(1 + \frac{r}{n}\right)^n - 1 = \left(1 + \frac{0.06}{2}\right)^2 - 1 = 6.09\%$$

In Example 2.1, the annual interest rate is 6%. Assume that the interest is compounded annually, semi-annually quarterly, monthly and daily. The effective interest rates are provided in Table 2.1.

Table 2.1   Annual nominal interest rate and annual effective interest rate

annual nominal interest rate, $r/\%$	interest period	number of periods in a year, n	rate in interest period/%	annual effective interest rate, $i/\%$
6%	annual	1	6	6
	semi-annual	2	3	6.09
	quarterly	4	1.5	6.14
	monthly	12	0.5	6.17
	daily	365	0.016	6.18

2.2 Basis Point and Cash Flow Diagram

According to above calculations, the more the interest periods per year, the larger the gap between the nominal and effective interest rates. In economic evaluation, the effective interest rate should be calculated according to actual situations.

2.2.1 Basis point

Due to the time value of capital, the value of the same amount of capital is different over time. Therefore, the capital at different time cannot be directly compared or merged. There will be a variety of capital flows at different time (Costs and benefits). The cash flows including costs and benefits are converted to a certain time point in the same year before they can be combined and compared. This point in that year is called the basis point in calculation. Theoretically, the basis point can be selected at the beginning of any year during the calculation period. In economic evaluation, it is generally set at the beginning of the first year of the construction period. Also, it can be set at the start of the first year of the operation period, but this should be noted.

When calculate the time value of capital, the basis point is set at the beginning of the first year of the construction period, according to *Regulation for Economic Evaluation of Water Conservancy Construction Projects* (SL 72 -2013). The input and output except for the loan interest are assumed to be incurred and settled at the end of the year.

2.2.2 Cash Flow Diagram

The construction and operation of water conservancy and hydropower projects usually last a long period. For large - scale water conservancy projects, it may take 5 - 10 years for construction. Consequently, the capital investment and return would create during cash flows. To objectively evaluate the economic effect of projects (or technical schemes), the amount of cash outflow and inflow, and the time when the cash flow occurs should be considered.

To intuitively and clearly illustrate the annual costs and benefits of water conservancy projects, the cash flow diagram can be used (see Figure 2.1). Here, x, y coordinates are time and capital, respectively. Each step represents one year. The arrow length represents fund value at a certain proportion. The arrows up and down represent cash inflow (i. e., income) and outflow (i. e., expenditure), respectively. Generally, the beginning of the project is taken as the origin (base point), and the years on the right side of the origin are construction period, start - up period and operation period. In economic evaluation of water conservancy projects, cash inflows and outflows are both occurred and settled at the end of the year. It is easy to see that the end of the first year is the beginning of the second year.

The cash flow diagram is an instrumental tool in economic evaluation and scheme analysis, and contributes to get a good understanding (Figure 2.1). However, in practical problems, considering the number of cash flow transactions and the complexity of charts, cash flow statements are often used instead of the cash flow diagram.

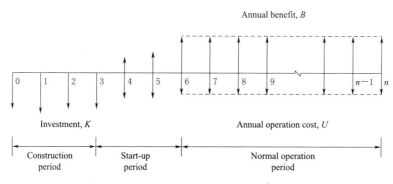

Figure 2.1 Cash flow diagram

2.3 Basic Formula of Dynamic Capital

Due to the time value of money, the earlier cash flows are more valuable than later cash flows. Therefore, the cash flows at different points over time cannot be directly added or subtracted. Instead, they need to be resolved into a set of cash flow equivalences. And this is usually done by compound interest method.

Some basic symbols used in the calculation are described as below:

P—present value of money. The principal or at the beginning of the year or principal converted to the beginning of the year;

F—final value of money. The future sum of money at the end of the nth interest period, which is equivalent to P with interest rate i;

A—equal annuity. The series of equal inflow or outflow of funds at the end of the years, from the 1st year to the nth year;

G—grade of the equal difference series;

i—interest rate, calculated interest rate or discount rate of banks;

n—number of interest periods or calculation periods, usually in years;

j—the growth percentage of the equal ratio series.

To facilitate equivalent computations, three types of interest formulas are derived, including: single payment formula, equal amount payment formula and arithmetic series formula.

2.3.1 Single Payment Compound Interest Formulas

The single payment formula is the lump-sum payment formula, which means that the cash flow, regardless of expenditure or income, occurs only once at a certain time. It is a one-time activity. A typical cash flow diagram is shown in Figure 2.2. According to different known conditions, there are two types of formulas, the lump sum future value formula and the lump sum present value formula.

1. Single payment compound interest formula

Suppose a given present value P, find the final value F after n years at interest rate i.

The interest at the end of the 1st year is iP. Then the-end-of-first-year total amount, i.e., the final value will be $F=P(1+i)$;

Figure 2.2 Typical cash flow diagram for single payment

The interest at the end of the 2nd year is $iP(1+i)$. Then the-end-of-second-year total amount, i.e., the final value will be $F=P(1+i)^2$;

The interest at the end of the 3rd year is $iP(1+i)^2$. Then the-end-of-third-year total amount, i.e., the final value will be $F=P(1+i)^3$;

Similarly, at the end of the nth year we will have: $F=P(1+i)^n$.

In the formula: $(1+i)^n$ is the final value coefficient of lump sum payment, or final value coefficient, designated as $(F/P, i, n)$.

Therefore, the final value formula for a single payment is as follows: $F=P(F/P, i, n)$

[**Example 2.2**] A water conservancy company needs to purchase survey equipment and loan 200,000 yuan from the bank at an annual interest rate of 10%. After 5 years, it will be settled and returned once. What is the amount (sum of principal and interest) to be repaid after 5 years?

Solution: Given $P=200000$ yuan, $i=10\%$, $n=5$ years

Then, the final value is calculated using the formula of one-time payment:
$$F=P(1+i)^n=200000\times(1+10\%)^5=322100 \text{ yuan}$$

That is, after 5 years, 322100 yuan need to be repaid.

In current economic evaluation, Excel spreadsheets are widely adopted by use of the Excel-internal functions. For example, Fv (Rate, Nper, Pmt, [Pv], [Type]) can be used to calculate the final value of a payment.

Where:

Rate—interest rate of each period;

Nper—total number of periods (calculation period is year, month, etc., the same as interest rate calculation period);

Pmt—the amount payable in each period, the value of which remains unchanged throughout the annuity period. In the calculation of the final value of a payment, the value is 0;

Pv—present value, or the cumulative sum of the current values of a series of future payments;

Type—0 or 1, used to specify whether the payment time is at the beginning or the end of the period. 0 means the payment time is at the end of the period. If [Type] omitted, the value is assumed to be 0.

Here in this example, enter "=Fv (10%, 5, 0, 20,)" in Excel, and the calculation result is 322100.

2. One-time payment present value formula (discount formula)

Given the final value after n years, F. The discount rate, i. The present value, P is calculated use the formula:

$$P = F/(1+i)^n$$

The present value formula of primary payment is the inverse operation of the final value formula of primary payment.

$1/(1+i)^n$ is the present value coefficient of primary payment, written as $(P/F, i, n)$.

Therefore, the final value formula of one-time payment is:

$$P = F(P/F, i, n)$$

[**Example 2.3**] Consider a project can earn 10000 yuan after 5 years, and the discount rate is 12%. What is the present value of its equivalent?

Solution: $P = F/(1+i)^n = 10000/(1+12\%)^5 = 5674$ yuan

That is 5674 yuan at present is equivalent to 10000 yuan five years later.

The Excel-internal function, Pv (Rate, Nper, Pmt, [Fv], [Type]) is used to calculate the primary payment present value. The parameters, Rate, Nper, Pmt, Fv and Type have the same meaning as the above example. Here in this example, enter "=Pv (12%, 5, 10000,)" in Excel, and the calculation result is 5674.

2.3.2 Equal Payment Formula

The calculation formula of equal payment consists of the formula of the final value of equal annuity, the formula of equal repayment capital, the formula of present value of equal annuity and the formula of capital recovery.

1. Formula of final value of equal annuity (annual value changes to final value)

Question: Given annuity value A, paid at the end of each year. What is the principal and interest (final value) F after n years?

The cash flow diagram is shown in Figure 2.3. Calculation details are as follows:

Repay A at the end of the first year, and get the final value F_1 at the end of the nth year;

Repay A at the end of the second year, and get the final value F_2 at the end of the nth year;

Repay A at the end of the $(n-1)^{th}$ year, and get the final value $F_{(n-1)}$ at the end of the nth year.

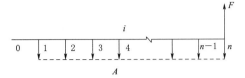

Figure 2.3 Formula of final value of equal annuity

Final value available at the end of the nth year:

$$F = F_1 + F_2 + F_3 + \ldots + F_n = A(1+i)^{n-1} + A(1+i)^{n-2} + \ldots + A(1+i) + A$$
$$F = A(1+i)^{n-1} + A(1+i)^{n-2} + \ldots + A(1+i) + A$$

Multiply $(1+i)$ on both sides of the above formula, then:

$$F(1+i) = A(1+i)^n + A(1+i)^{n-1} + A(1+i)^{n-2} \ldots + A(1+i)$$

Subtracting the two formulas, then:

$$Fi = A(1+i)^n - A$$

Namely,

$$F = A\left[\frac{(1+i)^n - 1}{i}\right]$$

In the formula, $\left[\dfrac{(1+i)^n - 1}{i}\right]$ is called uniform series compound amount factor or annuity final value factor. It is usually written as $[F/A, i, n]$.

[**Example 2.4**] Consider a water conservancy construction project invested by loans. The repayment method of equal principal and interest is adopted. At the end of each year, the repayment amount is 1 million yuan. Annual loan interest rate is $i=10\%$. How much is the accumulated repayment amount by the end of the 10th year?

Solution:

Given A=1 million yuan, $i=10\%$, $n=10$

$$F = A\left[\frac{(1+i)^n - 1}{i}\right] = 1000000 \times \left[\frac{(1+0.10)^{10} - 1}{0.10}\right] = 15937400 \text{ yuan}$$

The Excel-internal function Fv (Rate, Nper, Pmt, [Pv], [Type]) is used to calculate the final value of equal annuity. Here in this example, enter "= Fv (10%, 10, 1000000, 0, 0)" in Excel, and the calculation result is 15937400. All the parameters in Fv function have the same meaning as above.

2. Formula of equal repayment fund (fund deposit formula)

A capital amount of F need to be repaid after years. That is repay an amount of A at the end of each year during the n years. That is, a certain amount of A is deposit in advance at the end of each year during the n years. The accumulated money during n years can repay this capital. Therefore, it is also called fund deposit formula.

From $F = A\left[\dfrac{(1+i)^n - 1}{i}\right]$, the following equation can be derived:

2 Time Value of Capital and Equations for Cash Flow Equivalents

$$A = F\left[\frac{i}{(1+i)^n - 1}\right]$$

In the formula, $\left[\frac{i}{(1+i)^n - 1}\right]$ is called Sinking Fund Deposit Factor or Sinking Fund Factor. It is usually written as $[A/F, i, n]$.

[Example 2.5] A bank needs to repay a loan of $F = 1$ million yuan after 15 years. Then, what is the repaid capital, at the end of each year with an annual loan interest rat of 10%, during these 15 years?

Solution: Given $F = 1$ million yuan, $n = 15$, $i = 10\%$, then:

$$\begin{aligned} A &= F\left[\frac{i}{(1+i)^n - 1}\right] \\ &= 1000000 \times \left[\frac{0.1}{(1+0.1)^{15} - 1}\right] \\ &= 1000000 \times 0.0315 \\ &= 31500 \text{ yuan} \end{aligned}$$

The Excel-internal function PMT (Rate, Nper, Pv, [Fv], [Type]) is used to calculate the equal repayment capital. Enter "=PMT (10%, 15, 1000000, 0)" in Excel, and the calculation result is 31500. All the parameters in PMT function have the same meaning as above.

3. Present value formula of equal annuity

In scheme comparison of engineering projects, the annual benefits of a series of equal funds or the present value of annual cost are usually required. This is one of the discount methods. The economic significance of the present value formula of equal annuity is that: given compound interest rate i, and compound interest, calculate the present value P of equal payment cash A at the end of each period in n periods. That is, given A, i, n, find P.

Based on the final value of equal annuity $F = A\left[\frac{(1+i)^n - 1}{i}\right]$, and the final value of one-time payment $F = P(1+i)^n$, eliminate F, P is obtained as below:

$$P = A\left[\frac{(1+i)^n - 1}{i(1+i)^n}\right] = A(P/A, i, n)$$

In the formula, $\left[\frac{(1+i)^n - 1}{i(1+i)^n}\right]$ is called Uniform Series Present Worth Factor or Annuity Present Value Factor. It is usually written as $[P/A, i, n]$.

[Example 2.6] The annual generating revenue of a hydropower station is 120 million yuan. What is the total benefit present value with a social discount rate of 8% after a 50-year operation period?

Solution: Given $A = 120$ million yuan, $n = 50$, $i = 8\%$, then:

2.3 Basic Formula of Dynamic Capital

$$P = A\left[\frac{(1+i)^n - 1}{i(1+i)^n}\right]$$

$$= 120000000\left[\frac{(1+8\%)^{50} - 1}{8\%(1+8\%)^{50}}\right]$$

$$= 14.68 \times 10^8 \text{ yuan}$$

The Excel-internal function Pv (Rate, Nper, Pmt, [Fv], [Type]) is used to calculate the primary payment present value. All the parameters in Pv function have the same meaning as above. Here in this example, enter "=Pv (8%, 50, 1.2×10^8, 0, 0)" in Excel, and the calculation result is 14.68×10^8.

4. Fund recovery formula

An amount capital of P. The annual interest rate is i. Amortize the principal and interest at the end of each year during n years. That is, all the principal and interest shall be paid off after n years.

Repay principal and interest, A at the end of the first year, equivalent to the present value: $P_1 = \dfrac{A}{1+i}$;

Repayment of principal and interest, A at the end of the second year, equivalent to the present value: $P_2 = \dfrac{A}{(1+i)^2}$;

$\vdots$

Repayment of principal and interest, A at the end of the nth year, equivalent to the present value: $P_n = \dfrac{A}{(1+i)^n}$

$$P = P_1 + P_2 + \ldots + P_n = \frac{A}{(1+i)} + \frac{A}{(1+i)^2} + \ldots + \frac{A}{(1+i)^n}$$

That is,

$$P = A\left[\frac{1}{(1+i)} + \frac{1}{(1+i)^2} + \ldots + \frac{1}{(1+i)^n}\right] \tag{1}$$

Multiply $(1+i)^n$ at both sides of Formula (1), then:

$$P(1+i)^n = A\left[(1+i)^{n-1} + (1+i)^{n-2} + \ldots + 1\right] \tag{2}$$

Multiply $(1+i)$ at both sides of Formula (2), then:

$$P(1+i)^{n+1} = A\left[(1+i)^n + (1+i)^{n-1} + \ldots + (1+i)\right] \tag{3}$$

Substract Formula (2) from Formula (3), the fovlawing equatiton can be derived:

$$Pi(1+i)^n = A\left[(1+i)^n - 1\right]$$

Then:

$$A = P\left[\frac{i(1+i)^n}{(1+i)^n - 1}\right]$$

In this formula, $\left[\dfrac{i(1+i)^n}{(1+i)^n - 1}\right]$ is called Capital Recovery Factor or Fund Recovery

Factor. It is usually written as $[A/P, i, n]$.

[**Example 2.7**] A water conservancy project needs 10 million yuan for construction. All the money is loaned from a bank at the beginning of the year at one time. The annual loan interest rate i is 10%. Repayment starts when receiving the loan, and all the loan need to be repaid after 15 years. How much needs to be repaid at the end of each year?

Solution: Given $P=10$ million yuan, $i=10\%$, $n=15$, then:

$$A = P\left[\frac{i\,(1+i)^n}{(1+i)^n-1}\right]$$

$$= 10000000 \times \left[\frac{0.1\,(1+0.1)^{15}}{(1+0.1)^{15}-1}\right]$$

$$= 1314740 \text{ yuan}$$

In economic evaluation, the fund recovery formula is often used for the repayment of equal principal and interest. The Excel-internal function PMT (Rate, Nper, Pv, [Fv], [Type]) is used. Here in this example, enter " = PMT (10%, 15, 10000000, 0, 0)" in Excel, and the calculation result is 1314740. All the parameters in PMT function have the same meaning as above.

2.3.3 Formula of Arithmetic Series

The construction of water conservancy and hydropower projects usually takes a long time. Usually, with the progress of the project, the unit equipment increases year by year, the power generation benefit, annual operation cost also increase year by year, until all the units are installed. Simultaneously, the cash flow is an arithmetic series year by year. Next, capital equivalent calculation of this arithmetic series will be briefly introduced.

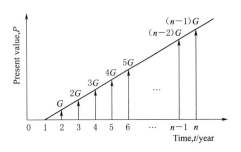

Figure 2.4 Cash flow diagram of arithmetic series

Assume an arithmetic series of cash flows: $0, G, 2G, \ldots, (n-1)G$ occurring at the end of $1^{st}, 2^{nd}, 3^{rd}, \ldots, n^{th}$ year. The annual interest rate is i. Calculate the period value F at the end of nth year, the present value P at the beginning of the first year and the annuity value A. See Figure 2.4 for the typical cash flow of differential series.

Generally, P occurs at the beginning of the first year, F occurs at the end of the nth year, and G occurs at the end of each year. It should be noted that this arithmetic series starts from zero, and the cash flow in the nth year is $(n-1)G$.

The calculating formulas for cash flow arithmetic series include the final value formula of the arithmetic series, the present value formula of the arithmetic series, and the annual value formula of the arithmetic series.

1. Final value formula of arithmetic series (Given G, n, i, Find F)

$$F = G(1+i)^{n-2} + 2G(1+i)^{n-3} + \ldots + (n-1)G \tag{1}$$

Multiply $(1+i)$ on both sides of Formula (1), then:

$$F(1+i) = G(1+i)^{n-1} + 2G(1+i)^{n-2} + \ldots + (n-1)G(1+i) \tag{2}$$

Subtract Formula (1) from Formula (2), then:

$$\begin{aligned}
Fi &= G(1+i)^{n-1} + G(1+i)^{n-2} + \ldots + G(1+i) - (n-1)G \\
&= G[(1+i) + \ldots + (1+i)^{n-2} + (1+i)^{n-1}] - (n-1)G \\
&= G\frac{(1+i)[1-(1+i)^{n-1}]}{1-(1+i)} - (n-1)G \\
&= G\frac{[(1+i)^n - 1 - i] - (n-1)i}{i} \\
&= G\left[\frac{(1+i)^n - 1}{i} - n\right]
\end{aligned}$$

Then,

$$F = \frac{G}{i}\left[\frac{(1+i)^n - 1}{i} - n\right]$$

In this formula, $\frac{1}{i}\left[\frac{(1+i)^n - 1}{i} - n\right]$ is called arithmetic series compound amount factor. It is usually written as $[F/G, i, n]$.

[Example 2.8] Consider a hydropower station with many units. The start-up period is 10 years. With the increase of the capacity of hydropower units year by year, the electricity revenue is an arithmetic series. Given that $G = 1$ million yuan, $i = 7\%$, $n = 10$ years. What is the present value of the total benefit of the hydropower station during the start-up period?

Solution: Given $G = 1$ million yuan, $i = 7\%$, $n = 10$ years, $A = 1$ million yuan

$$\begin{aligned}
F &= \frac{G}{i}\left[\frac{(1+i)^n - 1}{i} - n\right] + A\left[\frac{(1+i)^n - 1}{i}\right] \\
&= \frac{1000000}{7\%} \times \left[\frac{(1+7\%)^{10} - 1}{7\%} - 10\right] + 1000000 \times \left[\frac{(1+7\%)^{10} - 1}{7\%}\right] \\
&= 6833.7 \times 10^4 \text{ yuan}
\end{aligned}$$

Therefore, the present value of the total benefit of the hydropower station during the start-up period is 6833.7×10^4 yuan. See Figure 2.5 cash flow diagram of a hydropower station.

2. Present value formula of the arithmetic series (Given G, n, and i, find P)

Based on the final value formula of the arithmetic series and the formula $P = F/(1+i)^n$,

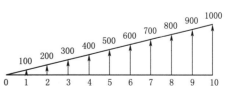

Figure 2.5 Cash flow diagram of a hydropower station

then:
$$P=\frac{G}{i(1+i)^n}\left[\frac{(1+i)^n-1}{i}-n\right]$$

In this formula, $\frac{1}{i}\frac{1}{(1+i)^n}\left[\frac{(1+i)^n-1}{i}-n\right]$ is called arithmetic series present worth factor. It is usually written as $[P/G, i, n]$.

3. Annual value formula of the arithmetic series

Bring the final value formula of the arithmetic series into the formula of equal repayment capital $A=F\left[\frac{i}{(1+i)^n-1}\right]$, then:

$$A=\frac{G}{i}\left[\frac{(1+i)^n-1}{i}-n\right]\left[\frac{i}{(1+i)^n-1}\right]$$

$$A=G\left[\frac{1}{i}-\frac{n}{(1+i)^n-1}\right]$$

In this formula, $\left[\frac{1}{i}-\frac{n}{(1+i)^n-1}\right]$ is called arithmetic series capital recovery factor. It is usually written as $[A/G, i, n]$.

[**Example 2.9**] A large irrigation area reconstruction project entered the trial operation period at the beginning of year 2000, and began to benefit. Due to the large areas, the initial operation period was as long as 10 years. It is estimated that the benefit will be 5 million yuan at the end of year 2000. With the increase of irrigation area, the annual benefit will increase by 1 million yuan. The annual interest rate is 6%. ①What is the present value at the beginning of year 2000 of the total benefit of in the initial operation period? ②What is the equivalent benefit value at the beginning of year 2010? ③How much is the equivalent benefit in each year?

Solution: Accordingly, the cash flow diagram is shown in Figure 2.6 (a). The unit of benefit is 10^4 yuan. The value of benefit at the beginning of year 2000 is marked as zero.

As this arithmetic series is different from standard arithmetic series, it must be divided into two parts: one is equal payment series and the other is standard increasing series. Specifically, the annuity value of the installment equal payment series is $A=5$ million yuan. The grade difference of the arithmetic series is $G=1$ million yuan. The cash flow diagram is shown in Figures 2.6 (a), (b) and (c).

Solution:

(1) The total benefit of the project in the initial operation period is equivalent to the present value at the beginning of year 2000. That is, the present value P_1 and P_2 of the benefits in Figures 2.7 (a) and (b).

$$P_1=A'\left[\frac{P}{A},i,n\right]$$

2.3 Basic Formula of Dynamic Capital

$$= A'\left[\frac{(1+i)^n - 1}{i(1+i)^n}\right]$$

$$= 500 \times \left[\frac{(1+0.06)^{10} - 1}{0.06(1+0.06)^{10}}\right] \times 10^4$$

$$= 500 \times 7.3601 \times 10^4 = 3680.05 \times 10^4 \text{ yuan}$$

$$P_2 = \frac{G}{i}\{[P/A, i, n] - n[P/F, i, n]\}$$

$$= \frac{100 \times 10^4}{0.06} \times \left[7.36009 - 10 \times \frac{1}{(1+0.06)^{10}}\right]$$

$$= 2960.22 \times 10^4 \text{ yuan}$$

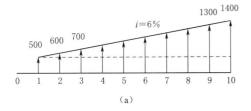
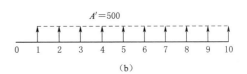

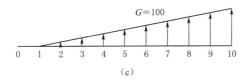

Figure 2.6 Cash flow diagram of a large irrigation area

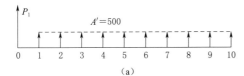
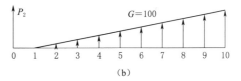

Figure 2.7 Cash flow diagram

Therefore, the total benefit equals to the present value P at the beginning of year 2000:

$$P = P_1 + P_2 = (3680.05 + 2960.22) \times 10^4 = 6640.27 \times 10^4 \text{ yuan}$$

(2) Given the present value P at the beginning of year 2000, actually find F:

$$F = P(1+i)^n = 6640.27 \times 10^4 \times (1+0.06)^{10} = 11891.7 \times 10^4 \text{ yuan}$$

(3) Given present value P, find the annual capital A.

$$A = P\left[\frac{i(1+i)^n}{(1+i)^n - 1}\right] = 6640.27 \times \left[\frac{0.06 \times (1+0.06)^{10}}{(1+0.06)^{10} - 1}\right] \times 10^4 = 902.20 \times 10^4 \text{ yuan}$$

2.3.4 A Summary of Basic Formulas For Dynamic Calculation

A summary of basic formulas commonly used in technical and economic evaluation is provided in Table 2.2.

Table 2.2   Basic formulas in dynamic calculation

Type	Name of Formula	known	unknown	Formula	Coefficient Name and Symbol
Single payment	Final Value Formula of Single Payment	P	F	$F=P(1+i)^n$	final value coefficient of lump sum payment $(F/P, i, n)$
	Present Value Formula of Single Payment	F	P	$P=F/(1+i)^n$	Present value coefficient of lump sum payment $(P/F, i, n)$
Equal Amount Payment	Final Value Formula of Equal Payment	A	F	$F=A\left[\dfrac{(1+i)^n-1}{i}\right]$	Uniform Series Compound Amount Factor $(F/A, i, n)$
	Equal Repayment Capital Formula (Fund Deposit Formula)	F	A	$A=F\left[\dfrac{i}{(1+i)^n-1}\right]$	Sinking Fund Deposit Factor $[A/F, i, n]$
	Present Value Formula of Equal Payment	A	P	$P=A\left[\dfrac{(1+i)^n-1}{i(1+i)^n}\right]$	Uniform Series Present Worth Factor $[P/A, i, n]$
	Capital Recovery Formula	P	A	$A=P\left[\dfrac{i(1+i)^n}{(1+i)^n-1}\right]$	Capital Recovery Factor $[A/P, i, n]$
Arithmetic Series	Final Value Formula of Arithmetic Series	G	F	$F=\dfrac{G}{i}\left[\dfrac{(1+i)^n-1}{i}-n\right]$	Final Value Factor of Arithmetic Series $[F/G, i, n]$
	Present Value Formula of Arithmetic Series	G	P	$P=\dfrac{G}{i(1+i)^n}\left[\dfrac{(1+i)^n-1}{i}-n\right]$	Present Value Factor of Arithmetic Series $[P/G, i, n]$
	Annual Value Formula of Arithmetic Series	G	A	$A=G\left[\dfrac{1}{i}-\dfrac{n}{(1+i)^n-1}\right]$	Amortization Factor of Arithmetic Series $[A/G, i, n]$

3

National Economic Evaluation

National economic evaluation is performed from the perspective of the whole society. It is regard - as economic cost - effectiveness analysis in *Construction Projects Economic Evaluation Method and Parameter (3rd Edition)*". The cost of investment activities is analyzed. The economic effects of project occupying economic resources, and resource allocation efficiency of project investment are evaluated. In national economic evaluation, shadow price, shadow exchange rate, shadow wage and social discount rate are used.

3.1 Requirements and Estimation Principles of Economic Benefit and Cost Identification

From the perspective of rational allocation of resources, national economic evaluation analyzes the economic efficiency of project investment and its contribution to social welfare. Also, it evaluates economic rationality of projects.

3.1.1 Requirements for Economic Benefits and Costs Identification

For water conservancy and hydropower construction projects, economic benefits and costs should be determined based on following requirements:

(1) Follow the "with and without comparison" principle.

(2) Analyze comprehensively the costs and benefits of all members and groups.

(3) Identify positive and negative external effects correctly. Avoid miscalculation, missing calculation or repeated calculation.

(4) Determine space scope and time span of benefits and costs, reasonably.

(5) Identify correctly the transfer payment, and adjust practical situations.

3.1.2 Calculation Principles for Economic Benefits and Costs

The principle of willingness to pay and/or willingness to accept compensation should be followed in calculating economic benefits. At the same time, the principle of opportunity cost should be followed in calculating economic cost.

(1) Principle of willingness to pay (WTP): follow this principle in calculating positive effects of project output. This is to analyze the value that social members are willing to pay for project benefits.

(2) Principle of willingness to accept (WTA): follow this principle in calculating negative effects of project output. This is to analyze the value of compensation received by social members for accepting adverse effects.

(3) Principle of opportunity cost: follow this principle in calculating economic cost of project input. This is to analyze opportunity cost of all resources occupied by the project. Opportunity cost is calculated according to the benefit of other most effective utilization of resources.

In national economic evaluation of water conservancy and hydropower construction projects, cost-benefit analysis determines the value of costs and benefits in monetary or qualitative terms.

3.1.3 Calculation Price of Economic Benefits and Costs

In national economic evaluation, shadow price is adopted for project inputs and outputs theoretically. Shadow price is used specifically as a calculation price in project economic analysis. It is determined according to pricing principle of economic analysis. Shadow price provides to understand real economic price of project inputs and outputs, relationship between market supply and demand, degree of resource scarcity, and requirements of rational allocation of resources. Without affecting the evaluation conclusions, shadow price can also be applied to the part of project benefits or costs which takes up a large proportion. For others, the financial price is suggested.

3.2 Cost Calculation

The costs of national economic evaluation are composed of direct and indirect costs. Omission and repetition should be avoided in calculating project costs.

Taxes and interest on domestic borrowing belong to internal transfer of national economy, so they should not be included in project costs.

For water conservancy and hydropower construction projects, the costs include construction investment (fixed-asset investment), operation cost (annual operation cost), working capital, maintenance operation investment (renewal and reconstruction cost) and external expenses. For projects with loans from foreign capital, the repayment of the principal and interest shall be paid project cash disbursement.

3.2.1 Construction Investment (Investment in Fixed Assets)

Construction investment in water conservancy construction project is also called fixed-asset investment. In national economic evaluation, construction investment (fixed-asset investment) includes all construction costs of main works and corresponding supporting works invested by the state, enterprises and individuals. Generally, the construction investment is estimated by budget estimation specialty according to construction period and construction scheme. Project capital is annually arranged for economic evaluation. Refer to *Section 4.2.1* for details of the construction investment.

In national economic evaluation, the internal economic transfer, i.e., the taxes (including construction tax, purchase tax of mechanical and electrical equipment, land acquisition tax, etc.) should be excluded.

Note that if the water supply benefit is calculated to the users, the investment in water distribution network and other supporting projects should be included according to the principle of consistent cost and benefit calculation caliber. The investment in supporting projects can be estimated by expanding typical design or referring to similar projects.

3.2.2 Operation Cost (Annual Operating Cost)

The operation cost (annual operation cost) includes all the operation costs during the initial and normal operation periods. They are calculated and adjusted according to project total cost. The depreciation cost, amortization cost, interest cost and water resource fee should be excluded from total cost as they belong to internal transfer. Refer to *Section 4.3* for details of cost estimation.

Some water supply projects have a long production period, and the annual operation costs can be determined based on production process, project scale and actual needs.

For foreign projects, the operation costs can be adjusted annually according to rising prices.

3.2.3 Working Capital

Working capital is the current capital occupied and used for a long time to maintain project operation during operation period. Specifically, it includes the purchase of fuel, materials, spare parts, and the payment of employees' wages. Refer to *Section 4.2.3* for working capital estimation.

3.2.4 Maintenance Investment (Investment in Renewal and transformation of Materials and Equipment)

Maintenance and operation investment (renewal and transformation cost) is the renewal and transformation cost of metal structure and mechanical and electrical equipment. It is determined according to project investment in metal structure and mechanical and electrical equipment. Generally, it is put into use before the expiration of the economic life of metal structure and mechanical and electrical equipment.

3.3 Benefit Calculation

3.3.1 Principles of Benefit calculation

According to *Regulation for Economic Evaluation of Water Conservancy Construction Projects* (*SL 72 -2013*), the national economic benefits of water conservancy projects should be calculated according to the following principles:

(1) Calculate project direct and indirect benefits through comparison before and after the project.

(2) The multi-year average annual benefits, as the basis of national economic evaluation, can be calculated by adopting the series method or frequency method. For projects serves for flood control, irrigation, urban water supply and others, the benefits of design year, super flood or drought year should be calculated for project decision-making study.

(3) The annual benefits during operation period should be calculated according to project production plan and matching degree.

(4) Take active measures to reduce the adverse effects of water conservancy construction projects on society, economy and environment. Otherwise, the negative benefits should be calculated.

(5) The residual of fixed assets and working capital of water conservancy construction projects shall be recovered once at the end of calculation period. Besides, project benefit flow should be considered.

(6) For multipurpose water conservancy construction projects, the benefit of each item and the overall benefit of the project should be calculated according to project function.

3.3.2 Benefits of Flood Control Project

Flood control projects are composed of reservoirs and dikes. The economic benefits of flood control projects have following features:

(1) Flood control projects cannot directly create wealth, but they provide safety services for the society and improve the living and production conditions of the people in the covered areas. Therefore, the reduced flood losses are taken as the benefits of flood control projects.

(2) Due to the random characteristic of floods, the benefits of flood control project vary greatly from year to year. Generally, there is no benefit in normal year. Only when

3.3 Benefit Calculation

the flood occurs can the value of flood control project be reflected. Therefore, the benefits of flood control are potential. Therefore, the benefits of flood control cannot be calculated based on the benefits of a certain year. Frequency analysis and statistical analysis of flood losses in each year are often needed.

(3) Flood losses include direct losses and indirect losses. Direct losses are caused by physical contact of the floodwater with humans, assets, property, and any other objects. Indirect losses are the subsequent or secondary results of the initial destruction. Specifically, the direct losses are composed of the loss of casualties, the loss caused by the damage of houses, facilities and materials, the loss caused by industrial and mining stop production, business suspension, disruption of traffic, power and communication interruption, the loss caused by production reduction or even no harvest in agriculture, forestry, animal husbandry, sideline and fishery industries, and expenses for flood control, emergency rescue and disaster relief. Comparatively, indirect costs are caused by illness, mental pain, return to normal production and daily life, etc. Some of the losses can be quantified in monetary terms (e.g., the area of flooded farmland, houses, machinery and equipment, disaster relief funds), while others are difficult to be quantified by money (e.g., illness, mental pain, etc.)

(4) With the development of national economy, in the flood control project protection area, the industry and agriculture are developing rapidly. At the same time, the population and property are increasing constantly. Therefore, even if flood occurs, the losses caused in different years are quite different. That is, the flood control benefits are changing with time.

(5) There are also some negative aspects in constructing flood control works. For instance, the construction of levees requires land acquisition and population relocation. The construction of flood detention reservoirs may inundate many lands, villages and towns. Besides, the construction of flood storage and detention areas needs to inundate land and villages during flood diversion. Negative benefit should be considered in flood control benefit.

(6) Flood control projects the financial burden can help lighten on flood control and disaster of national and local governments. Specifically, the flood control costs of rural, urban and traffic lines in the disaster area and affected areas will be reduced. The manpower, material resources of temporarily relocated residents in flood season, as well as the financial expenditures for disaster relief and rehabilitation will be also reduced.

Therefore, the flood control benefit refers to the reduced flood loss and the increased value of land development and utilization after flood control projects. It is usually in terms of the multi-year average benefit and the annual benefit of catastrophic flood. The loss of various protection objects after flood disaster can be calculated based on the inundation depth, inundation duration, and the situation of the protection area. The direct benefit of flood control (ice prevention) can be calculated by time series method and frequency curve method. The indirect benefit is generally estimated as a proportion of the direct benefit according to the survey data of typical years. Generally, the indirect benefit is 20%-30% of the direct benefit.

In the calculation of flood control (ice prevention) benefits, a lot of flood loss data is needed. Specifically, the basic data of flood loss should be investigated and analyzed as following:

1. Flood loss of urban and rural residents

It includes losses of houses, transportation, furniture and electrical appliances, family facilities, poultry and livestock raised in rural areas, etc. The property losses of typical families are investigated village by village and town by and town. Houses in typical villages and towns under different conditions, such as washing away, floating away, complete collapse and partial collapse, to determine the flooded loss, are investigated and analyzed.

2. Flooding losses of urban and rural enterprises

This can be divided into township enterprises, rural enterprises, urban industry, urban commerce, urban institutions and others. The fixed assets, working capital, semi-finished or finished products of enterprises and institutions shall be checked one by one.

For the loss value of urban and rural enterprises, it can also refer to the statistical value of inventory clearance before and after the disaster or the compensation value of insurance companies. Then, make supplementary investigation and adjustment to determine.

3. Loss of crops from flooding

The flood loss of crops usually last for one or two seasons or even multiple years.

(1) If the flood inundation only affects the crops in one season, but the crops cannot be replanted after the disaster, the loss shall be calculated by deducting the production and management costs according to the output value of the production reduction. If the flood is encountered, only the part of the production reduction shall be calculated. If the flood is followed by replanting, the loss is the difference between the output value of the replanting harvest and the output value of the normal year plus the increased production costs because of replanting.

(2) If the submergence time is too long and affecting the next season sowing, the loss shall be calculated according to the difference between the loss of two or more seasons and the normal year.

(3) Except for the property losses, if the floods have the benefits of fertilizing farmland and replenishing groundwater, these benefits will be deducted from the loss value of production reduction.

4. Loss of forest inundation

Forestry loss covers the loss of timber forest and economic forest. In general, the forestry loss is very small so it can be omitted. In flood storage and detention areas or crevasses with long inundation time and great flood, many trees may die. In the investigation, registration shall be made according to the tree species and tree age of various trees to calculate the forestry losses.

The flooding of fruit trees, tea trees and other economic forests may result in the reduction of output or the damage of trees in the same year. If the output of the current year is reduced, it is necessary to investigate the typical orchards and tea plantations affected by the disaster, analyze the output value of the affected year and the output value of the normal year, and calculate the flood loss. If the economic forest trees are damaged and need to be cut down and replanted, the flood loss shall be calculated as the loss for many years.

5. Flood loss of engineering facilities

The flood losses of irrigation and water conservancy facilities, highway bridges and culverts, rural power facilities and other engineering facilities can be calculated according to the ratio of the repair or reconstruction cost required to restore the damaged part of each department over the original situation and the pre-disaster engineering value. If the design standard for old facilities is updated during the restoration and reconstruction of engineering facilities, the added cost caused by the new standard should be substracted from the restoration cost.

6. Loss of special facilities

Special facilities refer to important public facilities such as railways, high-speed railways, expressways, high-voltage power grids, oil and gas pipelines. After the special facilities are damaged, the loss value includes the repair cost of special facilities and the loss caused by the interruption of special facilities.

The repair cost of special facilities to repair the broken or damaged is generally calculated accoeding to the damaged length, damage degree and cost per unit length. Considering the urgency, the repair cost of the special facilities will be higher than that under normal conditions. Therefore, a coefficient greater than 1 needs to be multiplied according to surveys.

[**Example 3.1**] The plain section of a high-speed railway has been destroyed for 3km. Under normal conditions, the investment per kilometer is 90 million yuan, and the emergency repair cost needs to be increased by 35% in flood period. Therefore, the required repair cost is: $9000 \times 3 \times 1.35 = 364.5$ million yuan.

The loss caused by the interruption of special facilities mainly refers to the reduced income during the interruption period. For example, the passenger and freight transportation revenue reduced by the interruption of railway and highway can be calculated according to the interruption days, passenger and freight traffic volume and the reduced net income; the reduced income by the interruption of power grid and oil and gas pipeline can be calculated according to the interruption duration, supply scope, supply population and the increased cost of alternative supply measures.

7. Other losses

Other losses refer to the expenses of rescuing the victims, transferring and resettling, relief materials and other expenses and unforeseen losses in flood control and emergency rescue. As the above-mentioned losses are difficult to be counted in flood disasters, the simplified method can be generally considered, and the estimation is based on 10%-20% of the total losses of the above types. The loss of a flood can be investigated and counted according to Table 3.1.

Table 3.1　　Investigation form of direct flood losses　　(　　Price level in year ×)

Items	Unit	Number	Value/10^4 yuan	Loss rate/%	Loss value/10^4 yuan
Ⅰ. Basic data					
1. Name of flooded region					
2. Area/km²					
3. Farmland/10^4 mu					
4. Population/10^4					
Non-agricultural people					
Agricultural people					
5. Mean ground elevation/m					
6. Inundation depth/m					
7. Submerge period/d					
8. Mean flow rate/(m/a)					
9. Transfer condition					
Ⅱ. Classification of direct flood losses					
1. Subtotal of crops					
1.1 Grains (rice, wheat, etc)					
1.2 Cotton					
1.3 Oilseeds					
1.4 Hemp					
1.5 Coarse cereals					
1.6 Vegetables					
1.7 Others					
1.8 Annual net production value					
2. Subtotal of forestry					

3.3 Benefit Calculation

Continued

Items	Unit	Number	Value/10^4 yuan	Loss rate/%	Loss value/10^4 yuan
2.1 Timber forest					
2.2 Shelter forest					
2.3 Fuel wood forest					
2.4 Economic forests					
2.5 Others					
2.6 Annual net production value					
3. Subtotal of fisheries					
3.1 Fish					
3.2 Intensive cultured aquatic product					
3.3 Other aquatic product					
3.4 Annual net					
4. Animal husbandry					
4.1 Cow					
4.2 Horse, mule colt, donkey					
4.3 Pig					
4.4 Sheep					
4.5 Chicken, duck, goose, rabbit					
4.6 Others					
4.7 Net production value					
5. Subtotal of engineering facilities					
5.1 Irrigation and water conservation					
5.2 Highway bridge and culvert					
5.3 Power supply facilities					
5.4 Telecommunication line					
5.5 Municipal shared facility					
5.6 Trough highway					
5.7 Trough railway					
5.8 Trough oil/gas pipeline					
5.9 Others					
6. Subtotal of people possess property					
6.1 Houses					
6.2 Production and transportation					
6.3 Furniture and home appliances					
6.4 Clothes					
6.5 Livestock and poultry					

Continued

Items	Unit	Number	Value/10^4 yuan	Loss rate/%	Loss value/10^4 yuan
6.6 Grain, grass and firewood					
6.7 Others					
7. Subtotal of enterprise property					
7.1 Fixed-asset capital					
7.1.1 Houses					
7.1.2 Facilities					
7.1.3 Others					
7.2 Working capital					
8. Subtotal of stop production of industrial and mining enterprises					
8.1 Annual net production value of industry and mining enterprises					
8.2 Annual profit and tax of business					
9. Sub-total of backbone transportation line interruption					
9.1 Net operation income of trough railway					
9.2 Net operating income of transit highway					
9.3 Net operating income of trough oil/gas pipeline					
9.4 Net income of electric power communication line					
10. Other subtotals					
10.1 Flood control and emergency rescue					
10.2 Victims rescue					
10.3 Resettlement of disaster victims					
10.4 Others					
11. Total					
Comprehensive index of direct flood loss Losses per mu: yuan/mu, losses per people: yuan/person					

The direct benefit of flood control can be calculated by time series method or frequency curve method. For long time series of disaster data, continuous 30 years or more is, the frequency curve method is suggested. Otherwise, if the data series method should be used to calculate the multi-year average of the short series, i.e., less than 30 years.

If the short series is used to calculate the multi-year average flood control benefits, special attention should be paid to the representativeness of the series. That is, the years of major disaster, small disaster and disaster free should be included.

(1) Time series method. Select a long series of flood disaster data and calculate direct flood disaster losses with and without flood control projects. Secondly, calculate the multi-year average annual losses by arithmetic average method. Then, the difference is the multi-year average direct benefits of flood control project. This method is simple and intuitive, but the disadvantage is that it is not easy to select the representative year series. If the series is too short, the representativeness is poor. Otherwise, more data are required if the series is long. Therefore, the data series should be selected according to actual situation in practice.

[**Example 3.2**] Consider a water conservancy project, located in the main stream of H River. The total reservoir capacity is 487 million m^3, and the installed capacity is 90MW. The project is multi-purposes, including ice prevention and power generation. The problem of ice jam in ×× section of H river is very serious. At present, under the condition of joint regulation of upstream reservoirs in the ice flood season, ice disaster of different degrees occurs almost every year during the ice flood period. A lot of manpower and material resources are invested in the rescue every year. According to statistics, the multi-year average annual direct economic losses caused by ice jam during 1992–2002 is 75.66 million yuan (calculated according to the comparable price in the first quarter of 2008, the same below). See Table 3.2 for the direct loss caused by ice jam in ×× river section.

Table 3.2   Direct loss caused by ice jam in ×× river section

Time period	Location of ice jam	Farmland /10^4 mu	Rooms	Population	Direct losses /10^4 yuan	Note
1992–1993	A	22.5		5597	5550	
1993–1994	B, C	35	1850	50000	5000	
1994–1995	B, D	19	358	1600	2147	
1995–1996	E	10.7	3061	7234	7760	
1997–1998	F, G	0.13	1708	8487	3826	A number of village and cities were affected
1998–1999	H			6308	1847	
2001–2002	B	1.82	1072		13522	
2002–2003	G	0.32			431	
Average					7566	

The indirect loss caused by ice jams is different in different ice jam disasters. Therefore, the indirect loss is estimated based on the investigation and analysis of typical ice jams. During 1993–1994, 1995–1996 and 2001–2002, the damage area and the influenced area is large, and the indirect loss is taken as 41% of the direct loss. During 1997–1998, the indirect loss is taken as 22% of the direct loss. In other small disaster years, the indirect loss is taken as 18% of the direct loss. Considering the indirect loss, the average losses caused by ice jams of ×× river section during 1992–2002 is 102.06 million yuan.

The reservoir regulating storage in the initial stage, 10 - year and 20 - year afterwards of the reservoir operation are 443 million m³, 207 million m³ and 127 million m³, respectively. The reservoir plays two important roles as follows:

1) Joint operated with the upstream reservoir during ice flood period, this reservoir will more effectively control the interval inflow, the diversion and retreat water in the irrigation area, and the channel water storage during the ice flood period. The discharge can be regulated timely and flexibly based on short - range weather forecast. This will make the downstream flow more regulated and balanced during the closure and opening period. Also, it will provide better dynamic conditions for the stable closure and opening of the river.

2) In winter, the power station releases water to generate electricity, causing the water temperature of the downstream channels increased. The river channel, several kilometers or even tens of kilometers downstream of the reservoir will become unsealed and unstable. Correspondingly, the ice amount will be reduced.

According to the ice jam analysis in typical years, the reservoir regulation and operation can firstly avoid low flow closure and maintain stable water level of the upper reaches during the river closure period and reduce the backwater level. Under the existing dike standards (or flood season protection standards), ice jam disasters with similar magnitudes of that in 1993, 1995 and 2001 will be avoided. Because of the ice reduction and uniform low discharge during the river opening period, the river discharge will be reduced by 127 million - 408 million m³. Consequently, the dynamic effect will be relieved so that even if an ice dam is formed in the lower reaches, the scale of the ice dam and backwater height of the reach will be greatly reduced. During river opening period, the maximum water level will be below the flood prevention level of the main flood season. This will significantly reduce the number of dike break disasters and losses.

Based on above analysis, during the ice flood period the reservoir regulation and operation can avoid ice jam loss in the upper reaches and lower reaches within 100km. The ice jam loss of the lower reaches can be reduced. Based on the price level of the first quarter of year 2008, the reduced ice jam loss at the lower reaches in the initial stage, 10 - year and 20 - year afterwards of the reservoir operation is 91.8 million yuan, 87.15 million yuan and 84.73 million yuan, respectively.

(2) Frequency curve method. Collect data on economic losses by different frequencies of floods to calculate flood control benefits. The calculation steps are as follows:

1) Calculate the disaster area and flood loss by different frequencies floods with and without flood control project, plot flood loss frequency curve with and without flood control project, as shown in Figure 3.1.

The two frequency curves in Figure 3.1 are both ladder shaped. This is because the river channels with and without flood control project, will be generally inundated only if the flood exceeds the safe discharge at a certain frequency. That is, the flood loss occurs suddenly, so the curve is ladder shaped.

2) The area enclosed by these two curves (abO and cbO) and the coordinate axes is the multi-year average annual flood loss with and without flood control projects. It can be calculated by graphic method.

3) The multi-year average annual flood control benefit is defined as the area difference of the two curves, i. e., abO and cbO. It can be calculated as follows:

$$S_0 = \sum_{p=0}^{p=1} (P_i - P_{i-1}) \left(\frac{S_i - S_{i-1}}{2} \right) = \sum_{p=0}^{p=1} \Delta P \, \overline{S}_{i,i-1}$$

Where:

S_0—annual average flood loss;

P_i, P_{i-1}—two adjacent frequency values;

S_i, S_{i-1}—flood losses at frequency of P_i, P_{i-1}.

In the calculation, the smaller the two adjacent frequency values are, the higher the calculation accuracy is. At the same time, the more data are needed.

[**Example 3.3**] The main tasks of a reservoir is located on the ×× river of ×× province, are industrial water supply, flood control, agricultural irrigation, ecology, etc. It is expected that, the flood control standard of the lower reaches will be raised. The return period (T) of future floods will be 30 years different from current 20 years. If a flood ($T \geq 30$ years)

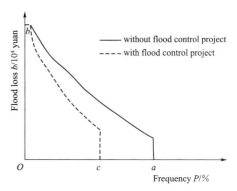

Figure 3.1 Flood loss frequency curve with and without flood control measure

occurs, the flood loss and inundation range of the region. In the reserve downstream of the reservoir, there are × county counties, high-speed railways, railways, expressways and national highways, 8 large and medium-sized coal mines, 210, 000 mu of farmland, 320, 000 people and a total area of 410 km². The basin protection zone is rich in industrial enterprises and mineral resources, especially coal resources. The total annual output of coal is more than 9 million tons.

(1) Calculate the reduced inundation area. According to the engineering situation of the reservoir, the flood routing is carried out for the design flood under various frequencies. The analysis results show that the construction of the reservoir can reduce the inundated area of the protection area in different degrees in case of various frequency floods. The flood return periods are classified to seven levels: 30 years, 50 years, 100 years, 500 years, 1000 years, 2000 years and 5000 years. The inundated area and loss rate of the project under different magnitudes of floods with and without projects area are analyzed respectively floods. The disaster reduction area and loss rate of the project are calculated

by "whether there is a comparison".

The farmlands in this area will be inundated by potential floods ($T \geqslant 30$ years) without flood control project. Comparatively, with flood control project, the flood control standard of the lower reaches of the reservoir will be raised from the current $T=20$ years to $T=30$ years. Accordingly, 210000 mu of farmlands in this area will be protected from floods. That is, the flood reduction area of this standard flood ($T \geqslant 30$ years) is 210000 mu.

(2) Calculate the direct flood loss. When direct flood loss First calculate the unit flood loss index. It is the total asset loss per unit area (mu) (yuan/mu) in the flooded area. The affecting factors include the asset value (in yuan/mu), flood loss rate (%) and flood loss growth rate.

Based on local national economic statistics and economic development, the calculated asset value of the reserve converted into farmland is 19500 yuan/mu.

The loss rate of flood disaster is the degree of asset loss in the project protection area. It is related to the topography, economic development, asset structure, submergence depth and submergence time. Based on the statistical data in 1993, the flood loss rate is estimated. The asset loss of reservoir protection area is shown in Table 3.3 based on existing data of preview projects.

Table 3.3 The asset loss of reservoir protection area (The average asset value is 19500 yuan/mu)

Flood returning period/year	Without flood control project		With flood control project	
	Average loss rate/%	Loss value/(yuan/mu)	Average loss rate/%	Loss value/(yuan/mu)
30	65	13000	0	0
50	70	14000	25	5000
100	75	15000	45	8775
500	80	16000	70	13650
1000	85	17000	75	14625
2000	90	18000	85	16575
5000	90	18000	90	17550

Based on above indexes, calculate the multi-year average annual asset loss in the reference year by frequency method. The average annual asset loss is 125.83 million yuan, and 46.86 million yuan without and with the flood control project, respectively. The difference is 78.97 million yuan. That is, the multi-year average annual direct flood control benefit of this project is 78.97 million yuan. Details is provided in Table 3.4.

(3) Calculate indirect flood control benefits. The reduced or avoided indirect national economic loss by flood control projects is called indirect national economic benefit of flood control, i.e., indirect flood control benefits. It consists of two parts: first, the social and economic flood losses caused by the interruption of traffic, power supply, communication and shortage of raw materials, etc. in the flooded area. Second, the cost required by

3.3 Benefit Calculation

Table 3.4 Calculation of multi-year average annual flood loss with or without the flood control project

With (without) flood control project	Frequency $P/\%$	ΔP	Loss, S_i /10^8 yuan	$S_{ave}=(S_i+S_{i+1})/2$ /10^8 yuan	$\Delta P \cdot S_{ave}$ /10^4 yuan	Multi-year average annal loss/10^4 yuan $S_0=\Sigma \Delta P \times S_{ave}$
Without flood control project	5					
	3	0.02	27.82	13.91	2318	2318
	2	0.01	29.96	28.89	3852	6170
	1	0.01	32.10	31.03	3103	9273
	0.2	0.008	34.24	33.17	2654	11927
	0.1	0.001	36.38	35.31	353	12280
	0.05	0.0005	38.52	37.45	187	12467
	0.02	0.0003	38.52	38.52	116	12583
With flood control project	3		0			
	2	0.01	10.70	5.35	713	713
	1	0.01	18.78	14.74	1474	2187
	0.2	0.008	29.21	23.99	1920	4107
	0.1	0.001	31.30	30.25	303	4409
	0.05	0.0005	35.47	33.38	167	4576
	0.02	0.0003	37.56	36.51	110	4686

enterprises and institutions to restore the production level, and the reduced benefit during the inundation period. Generally, the larger the flood, the larger the flood loss and the indirect loss. Based on existing statistical data and this project investment, the indirect flood loss is assumed as 20% of the direct loss. Accordingly, the calculated multi-year average annual indirect flood control benefit is 15.79 million yuan.

(4) Calculate (total) flood control benefits. Flood control benefit is the sum of direct and indirect flood control benefits. Herein, the multi-year average flood control benefit of this project is 94.76 million yuan.

3.3.3 Benefits of Waterlogging Control Project

The waterlogging control benefit of water conservancy construction project, usually in terms of multi-year average benefit and the benefit of extraordinary flood year, is calculated as the loss that can be reduced or avoided. Specifically, waterlogging losses are composed of four categories as below:

(1) losses caused by production reduction in agriculture, forestry, animal husbandry, sideline and fishery industries.

(2) losses caused by damage to houses, facilities and materials.

(3) losses caused by industrial and mining production stoppage, commercial shut-

down, traffic, power and communication interruption, etc.

(4) the costs for flood drainage and disaster relief.

Waterlogging loss includes direct loss and indirect loss. Direct loss is composed of crop production reduction, house damage, loss of means of production and damage of engineering buildings. Whereas, the indirect loss includes the loss in production and daily life in next year during post flood reconery. Indirect loss is usually calculated as some percentage of direct loss because it is difficult for quantification. The investigation of direct loss can refer to the flood loss investigation.

The benefits of waterlogging control can be calculated by frequency method or series method. If the frequency method is adopted, the waterlogging frequency method, waterlogging water volume method or rainfall – waterlogging correlation method can be used according to the characteristics and data of waterlogging area.

1. Waterlogging frequency method

This method is also called the actual annual series method. That is, based on the investigation data of waterlogging disasters, calculate the waterlogging loss from the established waterlogging loss frequency curves with and without the project. This method is suitable for long time series before and after the waterlogging control project. Based on the actual waterlogging areas, the multi – year average annual waterlogging area can be calculated. The difference between the average waterlogging area before and after the construction of waterlogging control project is the reduced waterlogging area. And the waterlogging control benefit is the product of the reduced waterlogging area and the loss rate of waterlogging per unit area. See Figure 3.2.

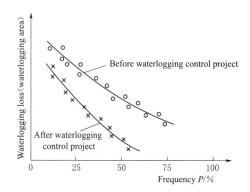

Figure 3.2 Waterlogging loss frequency curves

2. Waterlogging water volume method

In this method, first establish relation curve of waterlogging water volume and waterlogging loss based on measured and survey date. Next, calculate the waterlogging water volume and waterlogging loss before and after the waterlogging control project, and the difference is the waterlogging loss. This method is mainly used in plain polder area.

Waterlogging is affected by complex factors. The reduction of crop yield is closely related to waterlogging depth, ponding duration, groundwater level variation, crop varieties and crop growing period, etc. Under some conditions, the amount of waterlogging water can represent waterlogging depth, ponding duration and groundwater level variation. Therefore, we can start from the waterlogging amount and study the crop yield reduction rate. Next, calculate waterlogging loss.

(1) Following assumptions are made to calculate the waterlogging losses with and without the waterlogging control project:

1) the reduction rate β of agricultural yield changes with waterlogging water volume V, i.e., $\beta = f(V)$.

2) the waterlogging water volume V is a function of water level Z at the outlet of the waterlogged area, i.e., $V = f(Z)$. Assume that the waterlogging water volume changes with the water level of the control station and is not affected by the size of the river channel section.

3) assume that the disaster frequency is consistent with the precipitation frequency and the flow frequency at the control station.

(2) Calculate waterlogging loss by following steps:

1) According to the measured data of the hydrological station, the measured discharge hydrograph of the outlet control station in the waterlogging area before the flood control project is drawn.

2) The ideal discharge hydrograph is the assumed hydrograph when there is no waterlogging control project. That is, it is the discharge hydrograph when all drainage systems are unblocked without waterlogging. Calculate the peak discharge using runoff formula at small watershed or using the drainage modulus formula. Then, we can obtain the ideal flow hydrograph with generalized formula analysis, under comprehensive consideration of local topography and landform conditions.

3) Calculate the amount of waterlogging water. Compare the observed and ideal discharge hydrographs over the years, and the corresponding amount of waterlogging water can be obtained. As shown in Figure 3.3, curve I represents the ideal discharge hydrograph, and curve II represents the observed discharge hydrograph. The shadow area between the two curves is the waterlogging water amount over the years. Quantitatively, it can be calculated as follows:

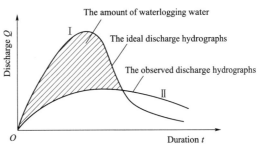

Figure 3.3 The observed and ideal discharge hydrograph

$$Q_i + Q_{i+1} - q_i - q_{i+1} = \frac{2 V_{i+1}}{t} - \frac{2 V_i}{t}$$

Where:

Q_i, Q_{i+1}—ideal discharge at time i and $(i+1)$, m³/s;

q_i, q_{i+1}—observed discharge at time i and $(i+1)$, m³/s;

V_i, V_{i+1}—waterlogging water amount at during time i and $(i+1)$, m³.

Start the calculation at the time i when ponding occurs, with $Q_i = q_i$, $V_i = 0$. Then, calculate V_{i+1} by above formula, and find out the corresponding water level from the reservoir capacity curve of the waterlogged area. In this way, the ponding water amount and the water level hydrograph in the waterlogged area can be obtained. Finally, calculate the maximum waterlogging water amount and maximum water level in the waterlogged area.

Calculate waterlogging area, flooding depth and flooding duration based on the waterlogging water amount, waterlogging depth and the elevation of farmlands.

4) Calculate crop yield reduction. Calculate the waterlogging depth and ponding duration of various crops before and after the construction of waterlogging control project. Then, find out crop yield reduction rate and yield reduction of various crops in Table of Crop Yield Reduction Rate by Waterlogging.

5) Calculate the benefit of controlling crop waterlogging. The benefits of controlling waterlogging of various crops in each year are calculated as the products of the yield reduction and the price of crop products before and after the waterlogging project construction. According to the requirements of waterlogging control engineering design stage, the waterlogging control benefits of typical years or long series of waterlogging control benefits can be calculated respectively as the basis for economic evaluation of waterlogging control projects.

3. Rainfall-waterlogging correlation method

First, investigate the relation between rainstorms and waterlogging losses based on historical waterlogging data. Before the construction of waterlogging control project. Next, calculate the waterlogging loss after the construction of the waterlogging project by using the storm frequency curve.

(1) In this method, following assumptions are made:

1) The rainfall less than or equal to the engineering treatment standard does not cause waterlogging.

2) The frequency of rainfall corresponds to that of waterlogging.

3) In the case of different design standards, the increase and decrease of rainfall is consistent with the increase and decrease of yield reduction rate under waterlogging.

(2) Calculation steps of the rainfall-waterlogging correlation method are as follows:

1) Investigate the waterlogging area and yield reduction rate of different flooded areas in each year before the construction of the project. Calculate the yield reduction rate of each year according to the following formula:

$$\alpha = \frac{\alpha_1 F_1 + \alpha_2 F_2 + \alpha_3 F_3 + F_4}{F}$$

Where:

α_1, α_2, α_3 —the yield reduction rate of mild, moderate and severe flood disasters;

F_1, F_2, F_3, F_4 —the waterlogged area and barren area suffered from mild, moderate and severe flood disasters;

3.3 Benefit Calculation

F—farmland area in waterlogged area.

The waterlogging area can be obtained from the flood control data of the water conservancy sectors or the statistical data of the national statistical sectors.

2) Draw the curve: $P+Pa-\beta$ as shown in Figure 3.4. β is the crop yield reduction rate by water logging. $(P+Pa)$ is rainfall (P: rainfall, Pa: antecedent precipitation) in the calculation period. Convert the figure into the first quadrant of Figure 3.5.

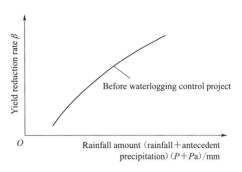

Figure 3.4 Relation curve of (rainfall+antecedent precipitation) - yield reduction rate

3) Select rainfall of different rainfall duration. For example, 1 day, 3 days, 7 days, 15 days, ..., and 30 days. Investigate the relationship between the selected rainfall and the waterlogging area or crop yield reduction rate. When the relation is good, that period is selected as the calculation rainfall period. Draw the rainfall frequency curve in the calculation rainfall period, as shown in the second quadrant of Figure 3.5.

4) Based on the rainfall frequency curve, $P+Pa-\beta$ curve, draw the frequency curve of waterlogging yield reduction rate before waterlogging control project by using the method of coaxial coherent chart. See the fourth quadrant of Figure 3.5.

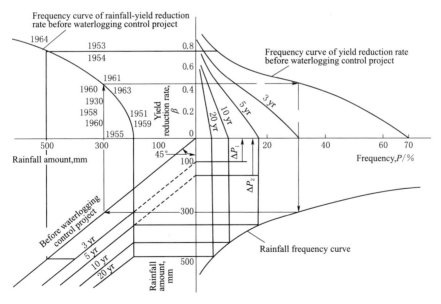

Figure 3.5 Coaxial coherent chart

5) According to the above assumption, after the waterlogging control project, waterlogging will occur only when the rainfall is greater than $(P+Pa)$. For example, after the

waterlogging control project, the rainfall amount with 3 year return period and 5 year return period will be increased by ΔP_1 and ΔP_2, respectively. The corresponding reduced waterlogging loss can be obtained in this way: draw two parallel lines in the third quadrant of Figure 3.5, then the intercepts at the vertical coordinate, i.e., ΔP_1 and ΔP_2 are the reduced waterlogging losses.

6) Draw the frequency curves of yield reduction rate with 3 year return period and 5 year return period along the arrow direction in Figure 3.5.

7) Calculate the difference between the average annual crop yield reduction rate before and after the waterlogging control project. It is the area between the frequency curve of crop yield reduction rate and the two coordinate axes. Then, calculate the average annual benefit of waterlogging control project.

The benefits of alkali control and waterlogging control projects should be analyzed and calculated based on the relationship between groundwater depth, soil salt content and crop yield. Also the projects' effect on reducing groundwater and soil salt content should be induded.

If the benefits of waterlogging control are closely related to the benefits of alkali control, then add up the benefits as the comprehensive benefits of the project.

3.3.4 Benefits of Irrigation Project

The economic benefits of irrigation projects are mainly reflected in the increase of output value caused by the improvement of crop yield or quality before and after irrigation or after the transformation of existing irrigated land. For example, the irrigation guarantee rate of the existing planting area in the irrigation area can be increased so as to realize the increase of production and income, the adjustment of crop planting structure (such as the conversion of dry land to paddy field, change one quarter crops to two or more quarters crops) and the irrigation facilities The improvement of application results in the extension of high yield crops. The irrigation benefit of water conservancy construction project refers to the increased output value of agriculture, forestry, animal husbandry and other products than that of without irrigation.

The characteristics of irrigation benefit are as follows: ① the improvement of crop yield and quality is the result of water, seed, fertilizer, soil improvement, agricultural technology and management measures, so the irrigation benefit needs to be shared between water conservancy and agricultural departments; ②due to different hydrometeorological conditions in each year, the irrigation benefit is also different every year. It can be represented by the multi-year average annual irrigation benefit, designed annual benefit or irrigation benefit in extremely drought year.

The irrigation benefit can be calculated by the method of sharing coefficient, shadow water price or water shortage loss. Water saving irrigation projects can improve irrigation conditions and save water resources. Water saving benefits can be calculated according to the benefits of water saving for expanding irrigation area or for urban water supply.

1. Sharing coefficient method

In this method, the total value by irrigation and agricultural technical measures with and without irrigation projects are compared. The irrigation benefit is taken as the product of the increased (i. e., difference) total value and the sharing coefficient. It can be calculated using the following formula:

$$B = \varepsilon \left[\sum_{i=1}^{n} A_i (Y_i - Y_{0i}) V_i + \sum_{i=1}^{n} A_i (Y'_i - Y'_{0i}) V'_i \right]$$

Where:
B—the average annual irrigation benefit, 10^4 yuan;
ε—the sharing coefficient of irrigation benefits;
A_i—the irrigation area, 10^4 mu;
Y_i, Y_{0i}—crop yield before and after the completion of the project, mainly determined by investigation and analysis, kg/mu;
Y'_i, Y'_{0i}—the output of by-products before and after the completion of the project, mainly determined by investigation and analysis, kg/mu;
V_i, V'_i—unit price of crops and by-products, yuan/kg.

This method is suitable for the projects with obvious changes in planting structure and agricultural technical measures while implementing the irrigation project. The total yield increases with and without the project is the comprehensive result of the irrigation project and crop variety improvement, fertilizer consumption increase, cultivation technology, plant protection measures, etc. Therefore, the benefits should be reasonably shared. The sharing coefficient of irrigation benefits can be divided into different years according to crop types (low water year, normal water year and high-water year), and the multi-year average annual value can also be used according to the crop types. Refer to the investigation data in similar areas for specific values. As an integral part of agricultural projects, the comprehensive benefits of the project should include the combined benefits of irrigation and agricultural technical measures. In general, the agricultural technical measures will be different with and without irrigation projects, so the sharing coefficient of irrigation benefits is less than 1.0. In China, the sharing coefficient of irrigation benefits is generally 0.2-0.6. The sharing coefficient of irrigation benefits in arid areas is larger than that in rainfall-abundant southern areas. It can be calculated based on historical statistical data or empirical data.

1) Calculate based on historical statistics:

$$\varepsilon = \frac{1}{2} \left(\frac{Y_{after} - Y_{before}}{Y_{after+tech} - Y_{before}} + \frac{Y_{after+tech} - Y_{tech}}{Y_{after+tech} - Y_{before}} \right) = \frac{(Y_{after} - Y_{before}) + (Y_{after+tech} - Y_{tech})}{2 \times (Y_{after+tech} - Y_{before})}$$

Where:
Y_{before}—the average annual yield per unit area of crops in several years without irriga-

tion project, kg/mu;

Y_{after}—the annual average yield per unit area, kg/mu. With irrigation project, but without agricultural technical measures;

Y_{tech}—the annual average yield per unit area, kg/mu. With agricultural technical measures, but the water conservancy conditions remain unchanged;

$Y_{after+tech}$—the annual average yield per unit area, kg/mu. The agricultural technical measures and irrigation projects are both implemented.

2) Calculate based on experimental data:

$$\varepsilon_{after1} = \frac{(Y_{after} - Y_{before})}{(Y_{after+tech} - Y_{before})} \quad \text{or} \quad \varepsilon_{after2} = \frac{(Y_{after+tech} - Y_{tech})}{(Y_{after+tech} - Y_{before})}$$

$$\varepsilon_{tech1} = \frac{(Y_{after+tech} - Y_{after})}{(Y_{after+tech} - Y_{before})} \quad \text{or} \quad \varepsilon_{tech2} = \frac{(Y_{tech} - Y_{before})}{(Y_{after+tech} - Y_{before})}$$

$$\varepsilon_{after} = \frac{\varepsilon_{after1} + \varepsilon_{after2}}{2} + \varepsilon_{tech} = \frac{\varepsilon_{tech1} + \varepsilon_{tech2}}{2} = 1$$

Where:

Y_{before}—average annual crop yield per unit area, kg/mu. Without irrigation project, but basic dry land agricultural technical measures, the same as that of the local farmers, are implemented for crop growth;

Y_{after}—average annual crop yield per unit area, kg/mu. With irrigation projects to fully meet the water demand of crop growth, but only basic agricultural technical measures are implemented as the above;

Y_{tech}—average annual crop yield per unit area, kg/mu. Without irrigation project, but efficient agricultural technical measures are implemented to fully meet the requirements of fertilizer, plant protection and cultivation for crop growth;

$Y_{after+tech}$—average annual crop yield per unit area, kg/mu. With irrigation projects, and good agricultural technical measures are implemented to provide good fertilizer, plant protection and cultivation for crop growth.

[Example 3.4] Calculate the irrigation benefit of wheat yield at an experimental station in northern Hebei Province based on the field data from winter in 2002 to summer in 2003. The wheat yield per mu is: $Y_{after+tech} = 623.8$ Jin[①], $Y_{before} = 268.7$ Jin, $Y_{tech} = 422.3$ Jin, $Y_{after} = 388.4$ Jin. Calculate the sharing coefficient of irrigation benefit.

$$\varepsilon_{after1} = \frac{Y_{after} - Y_{before}}{Y_{after+tech} - Y_{before}} \quad \text{or} \quad \varepsilon_{after2} = \frac{Y_{after+tech} - Y_{tech}}{Y_{after+tech} - Y_{before}}$$

$$\varepsilon_{after1} = \frac{388.4 - 268.7}{623.8 - 268.7} = 0.337 \quad \varepsilon_{after2} = \frac{623.8 - 422.3}{623.8 - 268.7} = 0.567$$

$$\varepsilon_{after} = \frac{\varepsilon_{after1} + \varepsilon_{after2}}{2} = \frac{0.337 + 0.567}{2} = 0.452$$

---

① 1 Jin=0.5 kg。

The experimental station is located in the temperate and semi-arid climatic zone, in the northern part of Hebei Province. As there is little rain in winter and spring, so the calculated irrigation benefit sharing coefficient is relatively high.

[Example 3.5] A water conservancy project is located in rainfall belt in southern China. The main task of the project is irrigation. The average annual irrigation water supply is 59.2 million m^3. It can improve the irrigation conditions of 301300 mu (including 230300 mu of paddy field and 71000 mu of dry land). Also, it can adjust the planting structure of crops and improve the multi-cropping index. The irrigation area is mainly planted with grain crops such as middle-season rice, ratoon rice, wheat, corn, pea, sorghum and sweet potato, and economic crops such as potato, rape, peanut, vegetable and fruit. Under the current situation, the area of grain planting in the irrigation area is 480700 mu, and the average annual total grain output is 156900 tons. The planting area of cash crops is 68700 mu, with the average annual total output of 49600 tons. The comprehensive multi-cropping index is 1.823. After the completion of the project, the grain planting area can reach 555900 mu, and the average annual total grain output can reach 242800 tons. The planting area of economic crops is 198600 mu, and the average annual total output is 174200 tons. The comprehensive multi-cropping index can be increased from 1.823 to 2.504. The unit yield of the crops has been improved with different levels. Considering the sharing coefficient, the multi-year average annual crop yield benefit is calculated as 77.44 million yuan. See Table 3.5 for details.

There are two aspects in calculating irrigation benefits by sharing coefficient method: the first is the allocation of water conservancy facilities and agricultural technical measures. The allocation coefficient is 0.2-0.6 determined by the water effect in agricultural yield increase. The second is the internal allocation of water conservancy facilities. Therefore, the benefit sharing between water source project and canal system project is generally determined by the investment proportions of these two projects. In "Example 3.5", the investment in water source project and irrigation canal system project are included in the project investment, so only the first aspect needs to be considered in the benefit calculation.

2. Shadow water pricemethod

In this method, the irrigation benefit is taken as the product of the irrigation water supply quantity and the shadow water price. The shadow water price method is suitable for the areas where the shadow water price of local irrigation water resources has been studied, and reasonable results have been achieved. Refer to *Section 3.3.5* for details.

[Example 3.6] According to the data in "Example 3.5", the average annual irrigation water supply is 59.2 million m^3. If the shadow water price in this area is 0.3 yuan/m^3, the irrigation benefit is 17.76 million yuan.

Table 3.5 Calculation results of yield benefits in irrigation area

Grain variety		Grain crops								Economic crops					Total
		Middle-season rice	Ratooning rice	Wheat	Corn	Sweet potato	Pea	Sorghum	Potato	Rape	Peanut	Vegetable	Fruit		
Current year	Planting area /10⁴ mu	21.69	4.59	10	3.27	6.02	2.06	0.44	0.24	3.76	0.93	1.69	0.25		54.94
	Unit yield /(kg/mu)	496	143	191	340	158	123	105	289	153	122	2400	570		5090
	Total yield /10⁴ t	10.76	0.66	1.91	1.11	0.95	0.25	0.05	0.07	0.58	0.11	4.06	0.14		20.64
	Unit price /(yuan/kg)	1.1	1.1	1.4	1.2	0.8	2.4	1.5	0.7	2.2	3	0.8	2		
	Total output value/10⁴ yuan	11834	722	2674	1334	761	608	69	49	1266	340	3245	285		23187
Planning year	Planting area /10⁴ mu	20.7	12.49	8.17	6.5	2.14	4.16	1.43	2.93	11.78	1.17	2.22	1.76		75.45
	Unit yield /(kg/mu)	620	310	320	480	280	230	200	580	210	180	5000	1100		9510
	Total yield /10⁴ t	12.83	3.87	2.61	3.12	0.60	0.96	0.29	1.70	2.47	0.21	11.15	1.94		41.70
	Total output value/10⁴ yuan	14117	4259	3660	3744	479	2296	429	1190	5442	632	8880	3872		49001
Output value of yield increase/(10⁴ yuan/mu)		2283	3537	986	2410	−282	1688	360	1141	4177	291	5635	3587		25814
Allocation coefficient of water conservancy project		0.3	0.3	0.3	0.3	0.3	0.3	0.3	0.3	0.3	0.3	0.3	0.3		
Allocated benefits of water conservancy project/10⁴ yuan		685	1061	296	723	−84	506	108	342	1253	87	1691	1076		7744

Note: This project is located in the subtropical and semi-humid area in the south. According to the date of the local agricultural experiment station, the water conservancy project sharing coefficient is 0.3.

3. Water shortage losses method

In this method, the irrigation benefit is taken as the agricultural yield reduction which is caused by water shortage. Specifically, the irrigation benefit is calculated as the product of the difference of crop yield reduction coefficients with and without irrigation project, the irrigation area, and the normal output value per unit area. The formula is as follows:

$$B = (d_1 - d_2) A Y S P$$

Where:

B—multi-year average irrigation benefits, 10^4 yuan;

d_1, d_2—multi-year average crop yield reduction coefficients with and without irrigation project, respectively;

A—control irrigation area of the project, 10^4 mu;

Y—crop yield at unit area, kg/mu;

SP—shadow price of unit crop yield, yuan/kg.

[Example 3.7] According to the data in "Example 3.5", given the current irrigation area, keep the planting area and planting structure after the irrigation project. Calculate the irrigation benefit by the water shortage loss method, and the result is 32.2 million yuan. See details in Table 3.6.

Table 3.6   Calculation results of irrigation benefit

Item	Grain crops							Economic crops					Total
	Middle-season rice	Ratooning rice	Wheat	Corn	Sweet potato	Vetch	Chinese sorghum	Potato	Rape	Peanut	Vegetable	Fruit	
Planting area /10^4 mu	21.69	4.59	10	3.27	6.02	2.06	0.44	0.24	3.76	0.93	1.69	0.25	54.94
Unit yield /(kg/mu)	496	143	191	340	158	123	105	289	153	122	2400	570	
Unit price /(yuan/kg)	1.1	1.1	1.4	1.2	0.8	2.4	1.5	0.7	2.2	3	0.8	2	
Crop yield reduction coefficients without irrigation project, d_1	0.15	0.15	0.15	0.1	0.1	0.1	0.1	0.1	0.15	0.1	0.15	0.15	
Crop yield reduction coefficients with irrigation project, d_2	0.05	0.05	0.05	0.05	0.05	0.05	0.05	0.05	0.05	0.05	0.05	0.05	
Benefits /10^4 yuan	1183	72	267	67	38	30	3	2	127	17	324	29	2161

Note: the production reduction coefficient without irrigation project can be determined based on the annual rainfall and water resources. With irrigation project, the irrigation water demand in each guarantee rate year is met, and the crops are not short of water. If the irrigation cannot be met in the year exceeding the guarantee rate, the average annual production reduction coefficient in this example is calculated as 0.05.

3.3.5 Benefits of Urban Water Supply Project

The urban water supply benefit of water conservancy construction project is calculated as the project benefits of providing production and domestic water to urban industrial and mining enterprises and residents. It is expressed in terms of multi-year average benefit, design annual benefit and extreme drought year benefit.

The urban water supply increases gradually. If a long time is taken before reaching the designed water supply scale, the annual production rate at the initial stage needs to be estimated. Next, calculate the water supply benefit based on the water supply amount of that year.

The benefits of urban water supply can be calculated by the optimum equivalent effect substitution method, comprehensive substitution method, shadow water price method and sharing coefficient method. Urban water includes domestic water, industrial and mining enterprise-utilized water and environmental water. In order to simplify calculation, the calculation method of urban residents' domestic and environmental water use benefits can be the same as industrial water use. According to *"Regulation for Economic Evaluation of Water Conservation Construction Projects"* (SL 72-2013).

1. Optimum equivalent effect substitution method

In order to meet the needs of urban residents living and industrial production, a variety of water supply schemes and water-saving measures are required. The water supply scheme is composed of the development of local surface water and groundwater, the construction of regulation and storage projects, and inter basin water transfer. The water-saving measures are composed of the improvement of water reuse rate, sewage purification and water use technology, the reduction of water transmission loss and water consumption quota, etc. Project investment and annual operation cost are required by alternative water supply and water-saving schemes. However, these costs can be avoided urban water supply project. Therefore, the benefit of urban water supply is taken as the annual cost of the alternative engineering measures. Specifically, the optimum equivalent effect substitution method is suitable in water shortage areas with prominent contradiction between water supply and demand.

[Example 3.8] In order to supply water to an industrial park, a reservoir needs to be constructed. The annual water demand of the industrial park is 85.36 million m^3, so a medium-sized reservoir is needed. The investment in fixed assets is 1797.46 million yuan, including 1039.34 million yuan invested in project construction and 758.12 million yuan invested in resettlement land. The water supply head (from the reservoir to the industrial park) is 50 m. The annual water pumping cost is 6.3 million yuan, and the annual operation cost is 39.75 million yuan.

The industrial park is located on the right bank of the Yangtze River. For water supply service, if the reservoir scheme is not adopted, a pumping station can be constructed as an alternative. It is planned to construct the pumping station as an alternative project for reservoir water supply. The fixed assets investment of the pump station is 1463 million yuan, including 856 million yuan of project investment and 607 million yuan of land occupation investment of immigrants. The lifting head is 160m. The annual water pumping cost is 35.55 million yuan, and the annual operation cost is 69 million yuan. Therefore, the water supply benefit of the reservoir is taken as the sum of the investment in the alternative fixed assets (1463 million yuan), and the annual operation cost (69 million yuan).

2. Comprehensive substitution method

After the construction of the urban water supply project, the costs of equivalent alternative projects and water-saving measures can be saved. Therefore, in the comprehensive substitution method, the benefit of urban water supply is taken as the annual cost of the comprehensive alternative projects and water-saving measures.

3. Shadow water price method

In this method, the benefit of urban water supply is the product of the shadow water price and the net water supply in the project area. The shadow water price is the price of commercial water. This method is suitable for the areas where the shadow price of water resources has been studied.

Shadow water price can be calculated by cost decomposition method, consumer willingness to pay method and opportunity cost method.

(1) Cost decomposition method. The cost decomposition method is an important method to determine the shadow price of non-foreign trade goods. The decomposition cost of water supply is obtained by decomposing the marginal cost of water supply (the average cost is often used in practice) and adjusting the shadow price. Decomposition cost refers to the value of all social resources needed to be consumed in the process of water supply, including various inputs of materials, labor and land, as well as the expenses that should be shared by capital investment. All kinds of costs need to be recalculated with shadow price. The specific steps are as follows:

1) Determine the shadow price of important raw materials, fuel, power, wages and other inputs, and calculate the operating cost of water per cubic meter.

2) The investment in fixed assets shall be adjusted and equivalent calculated. According to the investment proportion of each year in the construction period, the investment amount of fixed assets in each year during the construction period is calculated, and the annual fixed assets investment amount is converted to the beginning of the production period with the following formula:

$$I_F = \sum_{t=1}^{n1} I_t (1+i_s)^{n1-t}$$

Where:

I_F—unit fixed assets investment from equivalent calculation to the beginning of production period, yuan;

I_t—unit fixed assets investment after adjustment in each year of construction period, yuan;

n_1—construction period, years;

i_s—social discount rate.

3) Calculate the fixed fund recovery cost M_F of a cubic meter of water, without considering the recovery of the residual value of fixed assets:

$$M_F = I_F (A/P, i_s, n_2)$$

When considering the recovery of the residual value of fixed assets, it is as follows:

$$M_F = (I_F - S_v) (A/P, i_s, n_2) + S_v i_s$$

Where:

S_v—calculate the residual value of fixed assets recovered at the end of the period, yuan;

n_2—production period, year.

4) Calculate the working capital recovery cost of water per cubic meter, M_W:

$$M_w = W i_s$$

Where:

W—working capital per cubic meter of water.

5) The operating costs per cubic meter of water in the financial cost may not be adjusted.

6) The decomposition cost of water supply project is the sum of M_F, M_W and operation costs. It can be used as the shadow water price.

It is worth mentioning that the shadow water price is calculated by decomposing the marginal cost of water supply in this region, but not calculated as the decomposed cost of a water supply project.

(2) Willingness to pay method. The willingness to pay method is the price that is willing to pay for obtaining water. In economic analysis, the shadow price is often measured by consumers' willingness to pay.

(3) Opportunity cost method. In a proposed project, some resources may be replaced by other alternative opportunities. And the maximum economic benefit obtained from these alternative opportunities are called the opportunity cost. For example, water is a kind of resource widely used in industry, agriculture, daily life and environmental protection. If water is used in environmental protection project, then it can no longer be used in industry, life or irrigation. Then, the opportunity cost of the water is the maximum net benefit that obtained from industrial production and agricultural production.

[**Example 3.9**] A water supply project supplies 4.35 million m³ of water to urban residents. Calculate the shadow water price by the opportunity cost method.

The rice export port price is 330 dollars/t, the average yield is 500 kg/mu. The opportunity cost of water price is as follows:

Rice port price is: 330×6.12 (current exchange rate) ×1.08 (shadow price conversion coefficient) =2181.2 yuan/t

The transportation cost is: 20×1.84=36.8 yuan/t (transportation distance is 500 km, freight is 20 yuan/t, conversion coefficient is 1.84)

Trade cost: (2181.2−36.8)/(1+6%)×6%=121.4 yuan/t (6% is trade tax)

The shadow price in producing area is: 2181.2−36.8−121.4=2023 yuan/t

The benefit per mu is: 2023×0.5=1011.5 yuan/mu

Assume that the cost per mu is 40% of the price: 1011.5×40%=404.6 yuan/mu

The net benefit per ton of rice is: 1011.5−404.6=606.9 yuan/mu

The irrigation quota of rice is 350m^3/mu, and the shadow water price is: 606.9÷355=1.71 yuan/m^3

The annual average water supply benefit calculated by shadow water price method is: 1.71×4.35=7.439 million yuan.

In the case that the state and the region have not explicitly promulgated the shadow water price, according to *Regulation for Economic Evaluation of Water Conservation Construction Projects* (SL 72 -2013), the acceptable water price of urban water users can be taken as the shadow water price calculated from the willingness – to – pay method.

4. Apportionment coefficient method

In this method, the urban water supply benefit is approximately estimated as the product of the production increase value of industrial and mining enterprises and the sharing coefficient of water supply benefit. The apportionment coefficient method is suitable for the water supply project after the scheme optimization. It is a commonly used method in calculating the benefits of urban industrial water supply projects. The output value can be calculated as the water use per 10000 yuan of GDP. That is, the output value per cubic meter of water is the reciprocal of the water use per 10000 yuan of GDP. The added value can also be calculated with the total output value and the net output value. The total output value can be calculated as the water use per 10000 yuan of GDP. That is, the total output value of per cubic meter of water is the reciprocal of the water use per 10000 yuan of GDP. The net output value rate is the ratio of the increased value over the total output value. According to the statistical yearbook of water receiving area, the net output value rate is generally taken as 20%-30%.

The apportionment coefficient is a comprehensive function reflecting the interrelations among various factors which affect the benefit of industrial water supply. It is related to many factors, such as the level of industrial development, the status of water supply projects, industrial structure, investment structure, price system, water – saving level, management level, etc. The allocation coefficient of industrial water supply benefits can be calculated by investment ratio, fixed assets ratio, cost ratio, occupation fund ratio and discount cost ratio.

(1) Investment ratio is the ratio of industrial water supply investment to industrial production investment (including industrial water supply investment). It is used for calculating the apportionment coefficient.

(2) The ratio of fixed assets is the ratio of fixed assets of industrial water supply to fixed assets of industrial production (including fixed assets of industrial water supply projects). It is used for calculating the apportionment coefficient.

(3) Cost ratio is the ratio of industrial water supply cost over industrial production cost (including industrial water supply cost). It is used for calculating the apportionment coefficient.

(4) The ratio of occupation funds is the ratio of the funds occupied by industrial water supply (fixed assets and working capital) to the funds occupied by industrial production (including industrial funds for occupation).

(5) The discount cost ratio is calculated according to the ratio of the annual converted cost of industrial water supply (the present value of investment and annual operation cost) to the annual depreciation cost of industrial production (including the annual converted cost of industrial water supply).

Among the above methods, the three methods such as the ratio of occupied funds and the ratio of discounted expenses are comprehensive approaches considering the influence of benefit sharing coefficient. However, there are many difficulties in practical application: the cost ratio method has more contents for cost calculation; the cost composition and calculation content of different products are different; the data collection workload is very heavy. In practice, the data collection the calculation by investment ratio method and fixed assets ratio method are easy. Therefore, the investment ratio or fixed assets ratio is often used in practical work.

In addition to water sources, urban water supply projects generally need corresponding water conveyance projects, water treatment plants and municipal pipe network construction to form a complete water supply system and deliver water to users. If the water conservancy construction project includes water source, water conveyance project, water treatment plant and municipal pipe network, the benefits calculated above are the engineering benefits; if the water conservancy construction project is only a water source project, its engineering benefits shall be apportioned among the benefits of urban water supply projects, and the apportionment method is the same as above.

[**Example 3.10**] The main tasks of a reservoir are industrial water supply, flood control, agricultural irrigation, ecology, etc. The average annual industrial water supply of the reservoir is 22 million m^3.

According to the status of water in industrial production, the industrial net output value is multiplied by the apportionment coefficient to calculate the water supply benefit.

$$B = \frac{1}{D}\rho\xi$$

Where:

B—benefit of water supply per cubic meter (yuan/m^3);

D—water quota of 10^4 yuan output value within the scope of water supply (m^3/10^4 yuan); According to "*Integrated Water Resources Planning of* ××" development and Reform Commission and ×× Water Resources Sector. In the design level year of 2020, the water demand of ×× industrial output value is 18.5m^3 per 10000 yuan;

ρ—Net output value ratio. According to the statistical yearbook of ×× City, the total output value of industries above the municipal scale is 32893.58 million yuan, the industrial added value is 8832.68 million yuan, and the accumulated depreciation is 2192.93 million yuan. The industrial net output value is temporarily estimated according to the industrial added value minus the accumulated depreciation, and the industrial net output value rate is about 20%;

ξ—Water supply benefit sharing coefficient. It reflects the status and role of water in industrial production. Considering the severe water shortage situation in the current water supply area, water has become the primary restricting factor in industrial production. Only when there is water can production be produced, and when there is water, there will be benefit. According to the proportion of urban water supply project investment and industrial enterprise investment (including urban water supply project) within the water supply scope of the project, after calculation, the industrial water supply benefit sharing coefficient of the project is 5%.

After calculation, the water supply benefit per cubic meter of water is 5.4 yuan/m^3, and the annual average water supply benefit is 118.8 million yuan. According to the investment proportion of the project, water plant and water supply pipe network, the water supply engineering coefficient shared by the project is 85%, and the annual average water supply benefit of the project is 100.98 million yuan.

3.3.6 Benefits of Water Supply for People and Livestock in Rural Areas

The rural domestic water supply benefit of water conservancy construction project is from providing water for human and livestock to the countryside. It can be calculated by the optimum equivalent effect substitution method, comprehensive substitution method and shadow water price method.

1. Optimum equivalent effect substitution method

In this method, the benefit of water supply for people and livestock in rural areas are taken as the annual cost of the optimum equivalent effect substation project or water-saving measures. The equivalent substitution of people and livestock in rural areas is the saving of labor, animal power, machinery and corresponding fuel and materials for water transportation. Essentially, it is the cost of taking other measures for rural water supply without the project. Based on the investigation of the existing water supply source, water intake distance and water transportation mode, the equivalent substitution of people and livestock can be calculated. Also, it can be calculated based on the medical and health care costs or reduced diseases that can be saved by improving water quality.

2. Comprehensive substitution method

In this method, the benefit of waters supply for people and livestock in rural areas is taken as the annual cost of comprehensive alternative project combining projects and water saving measures.

3. Shadow water price method

In this method, the water supply benefit of rural people and livestock is taken as the product of the net domestic water supply quantity and the shadow water price in the project area.

For existing projects, the available water sources can satisfy the water demand from rural people and livestock in normal years. In typical low water years, water transportation is needed for water supply, because the water level of the river and lakes drops significantly, etc. Therefore, a combination of various methods is often used to calculate the benefits of water supply for rural people and livestock.

[Example 3.11] Consider a water supply project, mainly serves for irrigation and water supply for rural people and livestock. The multi-year average annual gross water supply is 1.03 million m^3. The water supply at the water diversion outlet of the water transmission pipeline is 1 million m^3. The benefit of water supply can be calculated as follows:

(1) Shadow price of water supply in normal years. In project area, the current water supply price is about 1.0 – 1.5 yuan/m^3. Considering that the rural living standard and disposable income will be improved in the design level year 2025, the shadow price of drinking water for people and livestock is determined as 2.0 yuan/m^3 in 2025.

(2) Water supply cost in dry years. Because of severe droughts in recent years, especially during 2001 – 2009, it is extremely difficult for people and livestock to drink water. Therefore, the local government can only organize vehicles to deliver water to the rural areas. The cost of water transportation is high because of poor rural roads. At present, for the new type of drought resistant water delivery truck, the fuel consumption per 100 km is 20 L. The rated load capacity is 4.5 t. The current local diesel price is 7.33 yuan/L. Assuming that the water delivery distance is 60 km, the unit drought resistant water delivery cost is 39.09 yuan/m^3. Additionally, 55% labor cost and other costs are considered. Thereafter, water delivery cost is 60.6 yuan/m^3. From statistical data, there were

seven years of severe drought from 1981 – 2005. In addition, the severe drought during 2001 – 2009 made it extremely difficult for people and livestock to get water. 31% of the years during 1981 – 2011 have droughts. Droughts in the county are mainly spring droughts (from March to May) and summer droughts (June to August) and droughts duration lasts for about 50% of the annual water supply period. Therefore, the water supply cost in drought resistant year is the average value of water supply price in normal year and water supply cost in drought resistance period. The value is 31.3 yuan/m^3.

(3) Comprehensive water supply price. The comprehensive water supply price is used to calculate the benefit of water supply for human and livestock. The calculation formula is as follows:

$$S = s_1 f_1 + s_2 f_2$$

Where:

S—comprehensive water supply price (yuan/m^3);

s_1— water supply price in dry years (31.3 yuan/m^3);

f_1—proportion of dry years (31.25%);

s_2—water supply price in normal years (2.0 yuan/m^3);

f_2—proportion of normal years (68.75%).

The price of water supply for people and livestock in rural area is calculated as 11.16 yuan/m^3. And the benefit of water supply is 11.16 million yuan.

3.3.7 Benefits of Hydropower Projects

The hydropower benefits of water conservancy and hydropower construction projects mainly include power consumption benefits, capacity benefits and energy saving and emission reduction benefits. The power consumption benefits and capacity benefit can be calculated by the optimum equivalent effect substitution method and shadow price method. The energy saving, and emission reduction benefit can be calculated by replacing thermal power generation with on – grid electricity of hydropower stations and reducing pollutant emissions. The power generation of hydropower station should adopt the average generation of many years and be calculated by series method. The effective capacity and effective energy, which can be absorbed by the power grid after power and energy balance, are used for capacity and power quantity calculation. For cascade hydropower stations and cross river diversion stations, the benefits of other cascade hydropower stations can be increased or reduced due to the construction of hydropower stations at the same level.

1. Optimum equivalent effect substitution method

In this method, the power generation benefit of hydropower station is taken as the construction investment and annual operation cost of the optimum equivalent effect substitution power station. Under the same power and electricity conditions, several feasible alternatives are selected. And the optimum alternative is the scheme with the minimum annual cost. The power consumption rate of thermal power station is larger than that of hydropower station. Therefore, the installed capacity of the alternative thermal power station shall be larger than that of hydropower station under the same capacity. The ratio is called capacity coefficient, and the general value is 1.1. At the same time, the power generation of thermal power station shall be greater than that of hydropower station. This ratio is called electricity quantity coefficient, and the general value is 1.05.

The construction cost of thermal power station is estimated by expanded index method, and the investment estimation of alternative thermal power installed capacity and unit installed capacity of thermal power station is adopted. At present, the construction investment of unit installed capacity of large thermal power units is generally 4000 – 5000 yuan/kW. The alternative thermal power is taken as the product of effective capacity and capacity coefficient, installed capacity. The effective capacity of hydropower station adopts the result of power and electricity balance. The operation cost of thermal power station includes fixed operation cost (operating cost) and fuel cost. The fixed operation cost (operating cost) is generally estimated by 4%–6% of the construction investment of thermal power station. The fuel cost is the product of the generation capacity of substitute thermal power plant, coal consumption per unit kW · h and standard coal price. The generation capacity of substitute thermal power station is the product of effective power generation capacity and electricity coefficient of hydropower station. In order to balance the electricity absorbed by the power grid, the maximum coal consumption of thermal power station is adopted as the coal consumption per unit electrical degree, and the coal price of the power grid is adopted as the standard coal price.

[Example 3.12] Consider a hydropower station with the installed capacity as 150 MW, and the average annual generating capacity as 590 million kW · h. According to the balances of electric power and energy, the power generation and peak regulation capacity in design – level year can be absorbed by the power grid. Calculate the power generation benefit based on the cost of thermal power station (the optimum equivalent effect substitution power station). The capacity coefficient and electricity coefficient of the alternative thermal power station is 1.1 and 1.05, respectively. The main parameters of the alternative thermal power station are as follows:

The investment per kilowatt: 4500 yuan/kW;

Coal consumption per kilowatt hour: 350 g/(kW · h);

Standard coal price: 360 yuan/t;

Fixed operating rate: 4%.

For the alternative thermal power station, the estimated capacity is 165 MW, and the generated energy is 619.5 million kW · h. The investment is 742.5 million yuan. The operation cost is 107.76 million yuan, including 29.7 million yuan of fixed operation cost and 78.06 million yuan of fuel cost. Therefore, the benefit of this hydropower station is

3.3 Benefit Calculation

taken as the sum of the investment (i. e. , 742. 5 million yuan) and operation cost (i. e. , 107. 76 million yuan) of the alternative thermal power station.

2. Shadow price method

In this method, the power generation benefit is taken as the product of the project-provided effective electricity quantity and the shadow price. According to *Regulation for Economic Evaluation of Water Conservancy Construction Projects* (SL 72 - 2013), the shadow price can be determined by calculating the marginal cost of power generation. Besides, the shadow price can also be calculated by decomposition cost method according to *Specification on Economic Evaluation of Hydropower Project* (DL/T 5441 -2010).

The cost decomposition method is an important method to determine the shadow price of non - foreign trade goods. The decomposition cost of water supply is obtained by decomposing the marginal cost of water supply (the average cost is often used in practice) and adjusting the shadow price. Decomposition cost is the value of all social resources consumed in water supply, including various inputs of materials, labor and land, as well as the expenses that should be shared by capital investment. All costs need to be re - calculated in terms of shadow price. Specifically, the calculation steps are as follows:

1) Determine the shadow price of important raw materials, fuel, power, wages and other inputs, and calculate the operating cost of 1 kW • h.

2) Adjust the construction investment and calculate the equivalent. Calculate the construction investment amount of each year based on the investment proportion of each year during the construction period. Calculate the equivalence of the annual construction investment amount to the beginning of the production period with the following formula:

$$I_F = \sum_{t=1}^{n_1} I_t (1+i_s)^{n_1-t}$$

Where:

I_F—unit construction investment from equivalent calculation to the beginning of production period, yuan;

I_t—unit construction investment after adjustment in each year of construction period, yuan;

n_1—construction period, years;

i_s—social discount rate.

3) Calculate the fixed capital recovery cost M_F of 1 kW • h without considering the recovery of residual value of fixed assets:

$$M_F = I_F (A/P, i_s, n_2)$$

When considering the recovery of the residual value of fixed assets, M_F is calculated as follows:

$$M_F = (I_F - S_V)(A/P, i_s, n_2) + S_V i_s$$

Where:

3 National Economic Evaluation

S_V—the residual value of fixed assets recovered at the end of the calculation period, yuan;

n_2—calculation period, year.

4) Calculate the working capital recovery cost M_w of 1 kW·h:

$$M_w = W i_s$$

Where:

W—working capital occupied by 1 kW·h.

5) In financial cost, the operation cost of 1 kW·h can keep un-adjusted.

6) The decomposition cost of the power station is the sum of M_F, M_w and operation cost. The decomposition cost can be used as the shadow price.

As the calculation of shadow price is complex, so shadow price is generally determined and released by the state investment authorities.

[Example 3.13] According to the data in "Example 3.12", the installed capacity of the hydropower station is 150 MW. The multi-year average power generation is 590 million kW·h. The power station has no peak load regulation capacity, and it operates at the base load in the power grid.

The shadow price in this area is 0.3 yuan/kW·h. The benefit of the power station is the product of the multi-year average annual generated energy and the shadow price. The estimated annual benefit of this project is 177 million yuan.

3. Benefits of energy saving and emission reduction

When the hydropower station is in operation, it will replace the thermal power generation, and reduce the environmental pollution. Specifically, it will reduce the emissions of carbon dioxide (CO_2), sulfur dioxide (SO_2), smoke and nitrogen oxides (NO_x). The benefits of energy saving, and emission reduction can be used to calculate the emission reduction. And the cost of industrial treatment of pollutants is taken as the benefits of the project.

According to *Guidelines for Calculating Ecological Benefit on the Project of Small Hydropower Substituting for Fuel* (SL 593 – 2013), the emission reduction benefits hydropower can be calculated as follows:

$$W_2 = 1.05 EQ K_3 K_4; \quad W_3 = 1.05 EQ K_5; \quad W_4 = 1.05 EQ K_6; \quad W_5 = 1.05 EQ K_7$$

Where:

EQ—electric quantity, kW·h;

W_2—CO_2 emission reduction of substituting fuel residual power for thermal power generation, t;

W_3—SO_2 emission reduction in place of fuel residual power instead of thermal power generation, kg;

W_4—smoke and dust emission reduction of substituting fuel residual power for thermal power generation, kg;

W_5—NO_x emission reduction of substituting fuel residual power for thermal power generation, kg;

K_3—the value of CO_2 emission into the atmosphere by burning 1 t standard coal, t. Herein, $K_3 = 2.567$ t;

K_4—conversion coefficient of coal consumption per unit electricity, t/(kW·h). Herein, $K_4 = 0.342 \times 10^{-3}$ t/(kW·h);

K_5—SO_2 emission coefficient, kg/(kW·h). Herein, $K_5 = 8.03 \times 10^{-3}$ kg/(kW·h);

K_6—dust discharge coefficient, kg/(kW·h). Herein, $K_6 = 3.35 \times 10^{-3}$ kg/(kW·h);

K_7—NO_x emission coefficient, kg/(kW·h). Herein, $K_7 = 6.90 \times 10^{-3}$ kg/(kW·h).

[**Example 3.14**] According to the data in "Example 3.12", the installed capacity of the power station is 210 MW, and the multi-year average annual power generation is 697.3 million kW·h.

According to calculation, the power station can reduce the emission of CO_2 by 642000 t, SO_2 by 5879 t, smoke by 2453 t and NO_x by 5053 t. The current industrial treatment costs of the above pollution are 124 yuan/t, 600 yuan/t, 1500 yuan/t and 10000 yuan/t, respectively. The annual emission reduction benefit is 137.43 million yuan.

3.3.8 Shipping Benefits

For water conservancy and hydropower construction projects, the shipping benefits are composed of the benefits from the navigation conditions provided or improved by the project and the benefits of water supply and navigation. The "with and without" method, the optimum equivalent effect substitution method or the comprehensive substitution method are used.

1. "With and without" method

Compare the transportation costs, transportation efficiency and shipping quality with and without the projects. The difference of benefits is taken as the shipping benefits. Specifically, the shipping benefits include: ①saved freight from replacing highway or railway transportation. ②saved costs from replacing waterway regulation or comprehensive measures combining waterway regulation with construction of railway (highway). ③benefits from improving the transportation efficiency and the shipping quality. It can save transportation, transfer and loading and unloading costs by improving berthing conditions and shipping conditions in ports. ④benefits from shortening the time of passengers and goods on the way, shortening the time of ships stopping at port, and shortening the waiting time for berthing in tidal river. ⑤benefits from improving the shipping quality and reducing marine accidents.

(1) Calculate the saved freight by alternatine highway or railway transportation using the following formula:

$$B_1 = C_w L_w Q_n + C_z L_z Q_z + \frac{1}{2} C_m L_m Q_g - \left(Q_n + Q_z + \frac{1}{2} Q_g\right) C_y L_y$$

Where:

B_1—the benefit of saving operation cost, 10^4 yuan/a;

C_w, C_z, C_y—unit transportation costs of no project, original related line and with project, yuan/(t · km), yuan/(person · km);

L_w, L_z, L_y—the transportation distance without project, original related line and with project, km;

C_m, L_m—the minimum unit transportation cost [yuan/(t · km), yuan/(person · km)] and corresponding transportation distance, km;

Q_n, Q_z, Q_g—normal transportation volume, transfer transportation volume and induced transportation volume, 10^4 t/a and 10^4 person times/a.

(2) Calculate the benefit of shortening the travel time of passengers using the following formula:

$$B_{21} = \frac{1}{2}(T_n Q_{np} + T_z Q_{zp})b$$

Where:

B_{21}—the benefit of shortening the travel time of passengers, 10^4 yuan/a;

T_n, T_z—the time saved for normal passenger transport and transfer passenger transport, h/person;

Q_{np}, Q_{zp}—the number of production personnel in normal passenger traffic volume and transfer passenger volume, 10^4 person times/a;

b—unit time value of passengers (calculated by per capita national income), yuan/h.

(3) Calculate the benefit of shortening the transit time of goods using the following formula:

$$B_{22} = \frac{SPQ \, T_s i_s}{365 \times 24}$$

Where:

B_{22}—the benefit of shortening the time of goods in transit, 10^4 yuan/year;

SP—Shadow price of goods, yuan/t;

Q—transportation volume, 10^4 t/a;

T_s—shorten the transportation time when there are projects, h;

i_s—social discount rate.

When calculating this partial benefit, it is necessary to deduct from the transportation volume the goods, such as grain, which will not affect the normal reserve.

(4) Calculate the benefit of shortening the stay time of ships in port using following formula:

3.3 Benefit Calculation

$$B_{23} = C_{sf} T_{sf} q$$

Where:

B_{23}—The benefit of shortening the stay time of ships in port is 10^4 yuan/a;

C_{sf}—Daily maintenance cost of the ship, 10^4 yuan/(ship/d);

T_{sf}—The shortened stay time of the ship in the whole year, d;

q—number of vessels, vessels.

(5) The benefit of improving shipping quality and reducing average accidents can be calculated as follows:

$$B_3 = aQSP + P_{sh} M \Delta J + B_{31}$$

Where:

B_3—benefit of improving shipping quality, 10^4 yuan/a;

a—the reduction rate of shipping cargo damage when there are items;

SP—shadow price of goods, yuan/t;

Q—transportation volume, 10^4 t/a;

P_{sh}—the average loss of shipping accident is 10^4 yuan per time. Refer to the existing accident compensation and handling situation;

M—shipping traffic volume (converted to t · km);

ΔJ—the reduction rate of shipping accidents with projects, times/ (10^4 t · km);

B_{31}—the reduced cost in dealing with difficult navigation and rapid flow channel, 10^4 yuan/a.

2. optimum equivalent effect substitution method or comprehensive substitution method

In optimum equivalent effect substitution method or comprehensive substitution method, the shipping benefit is taken as the investment and annual cost of the optimum equivalent effect alternative project or comprehensive alternative project.

3.3.9 Fishery Benefits

The fishery benefit of water conservancy and hydropower construction project is the benefit from aquaculture in the project water area as weel as by other measures. It is calculated as the product of the output of aquatic products and the price of aquatic products, considering the sharing coefficient with aquatic measures.

After the reservoir impoundment, the fishery yield will be increased by times of the water surface area. On the other hand, the biomass of the reservoir will be higher than that of the natural river. That is, the fishing productivity will also be higher than that of the natural rivers. As a result, the fishing yield of the reservoir will be significantly increased. The economic benefits can be calculated as follows:

$$B_{\text{fishery}} = \sum_{i=1}^{n} (S_i f_i - S_{0i} f_{0i}) C_i$$

Where:

B_{fishery}—fishery benefits after the project construction, yuan/a;

S_{0i}, S_i—the water surface area before and after the project construction, mu;

f_{0i}, f_i—fishing productivity before and after the project construction, kg/(mu·a);

C_i—the price of a fish species, yuan/kg.

[Example 3.15] After the construction of ×× Reservoir, the aquaculture water area has increased from 12 mu to 37 mu. Before of this, the average water surface yield of tilapia per mu is 500 kg/a. Currently, the average water yield per mu increases to 700 kg/a. The fish price is 5 yuan/kg. Then, the fishery benefit after the reservoir construction is: $B = (37 \times 700 - 12 \times 500) \times 5 = 99500$ yuan.

Labor costs, bait and aquatic measures are required to produce the fishery benefits. The sharing coefficient of aquatic measures can be calculated by the formula:

$$\varepsilon = \frac{Y_{\text{aquatic measures}}}{Y_{\text{labor}} + Y_{\text{bait}} + Y_{\text{aquatic measures}}}$$

Where:

ε—the sharing coefficient of aquaculture measures;

Y_{labor}—Labor cost per unit area, yuan/mu;

Y_{bait}—Food expenditure per unit area, yuan/mu;

$Y_{\text{aquatic measures}}$—Unit area aquatic product measure expenditure, yuan/mu.

The calculated labor cost is 150 yuan/mu. The calculated bait expenditure is 420 yuan/mu. The calculated aquatic product measures are 700 yuan/mu. The apportionment coefficient of aquatic measures is calculated as 0.55.

Considering the apportionment coefficient of fishery measures, the fishery benefit is 54700 yuan.

3.3.10 Benefits of Soil and Water Conservation

The benefits of water conservancy measures in water and soil conservation construction projects can be calculated in combination with the measures of agriculture, forestry and animal husbandry, so as to reduce the loss of water, soil and fertilizer, increase the output value of local agricultural, forestry and animal husbandry products, and reduce the losses caused by sediment to river channels, reservoirs and other water conservancy projects.

In the economic evaluation of water conservancy and hydropower construction projects, soil and water conservation benefits are generally not calculated quantitatively, but only as social benefits after construction. If it is required, refer to two specifications as below for calculation.

(1) *Comprehensive Control of Soil and Water Conservation -Method of Benefit Calculation (GB/T 15774 -2008)* is mainly applied for calculating the benefits of comprehensive management of soil and water conservation in small watershed. Specifically, the benefits are composed of basic benefits (water and soil conservation), economic benefits, social

benefits and ecological benefits. Particularly, the economic, social and ecological benefits are generated based on the water and soil conservation benefits.

(2) *Guidelines for Calculating Ecological Benefit on the Project of Small Hydropower Substituting for Fuel* (SL 593 -2013) is mainly applicable to the analysis and calculation of ecological benefits of small hydropower replacing fuel projects. Also, other hydropower construction projects can use it to a reference. The benefits of soil and water conservation include soil conservation, nutrient accumulation, water conservation, and reduction of water and soil loss area.

[**Example 3.16**] Consider a small hydropower project replacing fuel in Southwest China, the installed capacity of the power station is 7 MW. The multi-year average annual power generation is 25.72 million kW·h. The installed capacity of substituting fuel is 3.5 MW, and that of substituting fuel is 12.86 million kW·h. There are 1595 households using small hydropower alternating fuel. The area of protected forest is 1276 hectares. Calculate the benefits of soil and water conservation of the hydropower station.

Solution: According to *Guidelines for Calculating Ecological Benefit on the Project of Small Hydropower Substituting for Fuel* (SL 593 -2013), soil and water conservation benefits include soil conservation, nutrient accumulation, water conservation, and soil erosion reduction.

(1) Soil conservation benefits. The formula of soil conservation benefit is:
$$G_s = AP_s(X_2 - X_1)$$
Where:

G_s—forest conservation soil amount, t/a;

A—forest protection area, hm^2;

P_s—proportion of forest area with an inclination above 5 degrees, %;

X_1—soil erosion modulus of forest land, $t/(hm^2 \cdot a)$;

X_2—soil erosion modulus without forest land, $t/(hm^2 \cdot a)$.

The power station is in Southwest China. The soil erosion modulus of forest land is 500 t/km^2, and that of non-forest land is 4500 t/km^2. The proportion of forest area with inclination above 5 degrees is 8%. After calculation, 4083 t soil is conserved.

(2) Ecological benefits of accumulated nutrients. The ecological benefits of accumulated nutrients include the reduction of soil salt loss by forest and the increase of soil nutrients by decomposition of forest litter, etc. The calculation formula is as follows:
$$G_n = NG_s; \quad G_p = PG_s; \quad G_k = KG_s$$
Where:

G_n, G_p, G_k—reduce the loss of nitrogen, phosphorus and potassium, t/a;

N, P, K—the content of nitrogen, phosphorus, and potassium in soil, %.

According to *the 1:4 million Soil Nutrient Map of China* produced by the Institute of Agricultural Resources and Regional Planning, Chinese Academy of Agricultural Sciences: the soil nitrogen content is 0.4%, the phosphorus content is 0.08%, and the potassium content is 2.5%. Based on the quantity of conserved soil, the accumulated nutrient benefit is composed of the reduction of nitrogen loss (16t), the reduction of phosphorus loss (3t) and the reduction of potassium loss (102t).

(3) Water conservation can produce ecological benefits including precipitation conservation, flood mitigation, water quality purification, and increase of available surface water. The calculation formula is as follows:

$$G_t = 10A(R - E_v - C)$$

Where:

G_t—forest regulated water volume, m^3/a;

R—rainfall, mm/a;

E_v—stand evapotranspiration, mm/a;

C—surface runoff, mm/a.

(4) The area of soil and water loss reduction should be calculated according to the difference of annual average land area before and after the implementation of the project. The land loss includes the area damaged by gully erosion and the area of land "petrified" and "decertified" by surface erosion.

3.3.11 Land Value – added Benefits

Due to the improvement of design standards for flood control and waterlogging drainage, the barren land around the project can be turned into farmland. The land that can only be used seasonally can be used all year round. The farmland that can only grow low-yield crops can be turned into high-yield crops. And the farmland originally used as agricultural cultivation is changed into urban and industrial land, thus increasing the value of land development and utilization.

Since the increased land development and utilization value is mainly reflected in the difference of net income created by different land uses, the increased land development and utilization value is calculated according to the difference between the net land income with and without projects.

The value – added benefit of agricultural land is equal to the increase of net income caused by the change of low – value crops to high – value crops. The unit area benefit can be regarded as the increment of crop output before and after the construction of the project multiplied by the product price, and the value – added benefit of engineering agricultural land can be calculated according to the area affected by the project and considering a certain sharing coefficient.

The urban land value – added benefits can use for reference the existing projects, compare the change of land transfer price before and after the project construction, calculate the land value – added benefit according to the area affected and radiated by the water con-

3.3 Benefit Calculation

servancy project, and calculate the water conservancy project sharing coefficient according to the proportion of water conservancy project investment in a series of engineering construction, so as to obtain the land value – added benefit of the project. It can be calculated according to the following formula:

$$B = (P_2 - P_1)A\xi$$

Where:

B—land value – added benefit, in 10^4 yuan/hm^2;

P_1, P_2—land transfer price before and after the project construction, in 10^4 yuan/hm^2;

A—land area affected by the project, in hm^2;

ξ—sharing (or allocation) coefficient of water conservancy project.

[Example 3.17] Consider a river, located in ×× City, and the river channel under river regulation is 24.651 km in length. Calculate the land value – added benefits of the construction project based on relevant data.

According to the statistical data of Land and Resources Bureau of ×× City, the land transfer price on both sides of the river after river regulation is increased, from 500 million yuan/km^2 to 1200 million yuan/km^2. See Table 3.7 for details of the river land value – added benefits.

Table 3.7   Statistical data of land use conditions in the surrounding area of the regulated river reaches of ×× City

Name	Reach length /km	Item	Developed area		Undeveloped area		Protection area
			Residential area	Others	Residential area	Others	
Reach A – B	7.42	length/km			7.42	7.42	7.42
		Width/km			0.14	0.28	0.39
		Area/km^2			1.05	2.09	2.87
Reach B – C	4.378	length/km			4.38	4.38	4.38
		Width/km			0.15	0.09	0.32
		Area/km^2			0.68	0.40	1.40
Reach C – D	5.073	length/km			5.07	5.07	5.07
		Width/km			0.40	0.43	0.18
		Area/km^2			2.03	2.16	0.93
Reach D – E	7.78	length/km	7.78		7.78	7.78	7.78
		Width/km	0.05		0.15	0.30	0.41
		Area/km^2	0.37		1.19	2.34	3.19
Total	24.651		0.37	0.000	4.95	6.99	8.39

According to Table 3.7, after the project, the undeveloped land area on both sides of the river is 11.94 km^2. The value – added per unit area is 700 million yuan/km^2. Therefore, the land value – added benefit is 8358 million yuan. It is generated by the comprehensive impacts of economic development, river regulation and municipal engineering. Conservatively, assume that the water conservancy projects share 25% of the land value-added benefits. Then, the value-added benefit of water conservancy projects is 2090 million yuan.

$$B=(P_2-P_1)A\xi=(12-5)\times 11.94\times 25\% =20.9\times 10^8 \text{ yuan}$$

Given the benefit period as 50 years, so the average annual benefit is 41.8 million yuan.

3.4 National Economic Evaluation Indicators

In national economic evaluation of water conservancy and hydropower construction projects, three indicators are used in economic cost-benefit analysis: economic internal rate of return (EIRR), economic net present value (ENPV) and economic benefit-cost ratio (R_{BC}).

(1) Economic internal rate of return (EIRR) is the discount rate when the cumulative present value of net benefits of each year during calculation period is equal to zero. The formula is as follows:

$$\sum_{t=1}^{n}(B-C)_t(1+\text{EIRR})^{-t}=0$$

Where:

EIRR—economic internal rate of return;

B—annual benefit, 10^4 yuan;

C—annual cost, including investment and annual operation cost, 10^4 yuan;

$(B-C)_t$—the net benefit of the tth year, 10^4 yuan;

t—the serial number of each year in the calculation period, for base point it is zero.

The economic rationality of projects can be judged by comparing EIRR with social discount rate i_s. The project is economically reasonable when $EIRR \geqslant i_s$. EIRR can be calculated after getting the net benefit during calculation period by using Excel-internal function IRR (values, [guess]).

(2) Economic net present value (ENPV) is the sum of the net benefits of each year in the calculation period discounted to the present value at social discount rate i_s. The formula is as follows:

$$ENPV=\sum_{i=1}^{n}(B-C)_t(1+i_s)^{-t}$$

Where:

$ENPV$—economic net present value, 10^4 yuan;

i_s—social discount rate.

The economic rationality of projects should be judged based on ENPV. The project is economically reasonable when ENPV ≥ 0. ENPV can be calculated by using Excel – internal function NPV (rate, value1, [value2],...).

(3) Economic benefit – cost ratio (R_{BC}) is the ratio of benefit present value over cost present value during calculation period. The formula is as follows:

$$R_{BC} = \frac{\sum_{i=1}^{n} B_t (1+i_s)^{-t}}{\sum_{i=1}^{n} C_t (1+i_s)^{-t}}$$

Where:

R_{BC}—economic benefit cost ratio, in 10^4 yuan;

B_t—the benefits of the tth year, in 10^4 yuan;

C_t—the expenses of the tth year, in 10^4 yuan.

The economic rationality of projects should be judged based on R_{BC}. The project is economically reasonable when $R_{BC} \geq 1.0$.

The benefits and costs during calculation period can be calculated using the Excel – internal function NPV (rate, value1, [value2],...). Then, EIRR can be obtained by dividing the benefit present value by the cost present value.

3.5 National Economic Evaluation

1. social discount rate

Social discount rate is a general parameter of national economic evaluation of construction projects, and it is the main basis for judging economic feasibility of construction projects. Social discount rate represents the minimum rate of return required by social investment. According to *Construction Projects Economic Evaluation Method and Parameter* (3rd Edition), the recommended social discount rate is 8%.

According to *Regulation for Economic Evaluation of Water Conservancy Construction Projects* (SL 72 -2013), the social discount rate should use the state – stipulated value of 8% in national economic evaluation. Water conservancy projects, e. g., flood control, drainage, ecological management projects, have unique functions of social public benefits including social benefits, environmental benefits, and other benefits. However, these benefits are difficult to reflect in currency. Therefore, the social discount rate at 6% can be used in economic evaluation.

2. Calculation Period

The calculation period in national economic evaluation of water conservancy and hydropower construction projects includes the construction period and operation period. The operation period is generally 50 years. Simultaneously, it is necessary to consider the maintenance and operation investment of mechanical and electrical equipment and metal structure (investment in renewal and transformation). Generally, the discounted base year is set at the beginning of the first year of the construction period.

3. Benefit – cost flow

In national economic evaluation, the net cash flow and related economic indicators are calculated based on the calculated benefit flow and cost flow. The cost benefit flow of water conservancy projects is shown in Table. 3. 8. The cost – benefit flow of hydropower projects is shown in Table 3. 9.

Table 3. 8　　　　　Cost – benefit flow (water conservancy projects)

Serial number	Item	Total	Construction Period			Operation Period		
			1	2	...	...	$n-1$	n
1	Benefit – cost flow							
1.1	Benefits of different project functions							
1.1.1	×××							
1.1.2	×××							
1.1.3	×××							
1.2	Recovery of the residual of fixed assets							
1.3	Recovery of working capital							
1.4	Indirect benefits of project							
1.5	Negative benefits of project							
2	Expense flow							
2.1	Fixed – asset investment							
2.2	Working capital							
2.3	Annual operation cost							
2.4	renewal and reconstruction cost							
2.5	Indirect costs of project							
3	Net benefit flow							
4	Accumulated net benefit flow							
Evaluation indicators: EIRR (　%); ENPV ($i_s=$　%); R_{BC} ($i_s=$　%)								

Table 3. 9　Economic Cost – Benefit Flow of Project Investment (Hydropower project)

Serial number	Items	Total	Construction Period			Operation Period		
			1	2	...	...	$n-1$	n
	Annual installed capacity/MW							
	Annual effective capacity/MW							
	Annual effective energy/(10^4 kW·h)							
1	Cost – benefit flow of alternative scheme							
1.1	Construction investment							
1.2	Maintenance operation investment							

3.5 National Economic Evaluation

Continued

Serial number	Items	Total	Construction Period			Operation Period		
			1	2	⋯	⋯	$n-1$	n
1.3	Operation cost							
1.4	Fuel cost							
1.5	Others							
2	Design cost flow							
2.1	Construction investment							
2.2	Maintenance operation investment							
2.3	Operation cost							
2.4	Working capital							
2.5	Project Indirect Costs							
3	Net Benefit Flow							
Calculation indicators	EIRR/%							
	ENPV ($i_s =$ %)							
	Benefit-cost ratio ($i_s =$ %)							

4

Financial Evaluation

Financial evaluation is performed from project perspective under current national financial and tax system and price system. Specifically, financial benefits and costs are calculated, and financial statements are prepared. Evaluation indicators are calculated, and the viability, solvency and profitability of projects are examined. All these are provided to judge project financial feasibility and provide basis for decision making.

4.1 Basic Principles, Price System and Evaluation Contents of Financial Evaluation

4.1.1 Basic Principles

The basic principles of financial evaluation include:
(1) consistency in the calculation of cost and benefit.
(2) principle of with-and-without in the identification of cost and benefit.
(3) combine dynamic analysis with static analysis, and take the dynamic analysis as the main.
(4) the sound principles established by basic data.

4.1.2 Price System

The financial evaluation of water conservancy and hydropower construction projects is based on the proposed fund sources and different financing plans, It adopts financial prices according to the current national financial and tax system.

According to *Regulation for Economic Evaluation of Water Conservancy Construction Projects* (SL 72 -2013) and *Specification on Economic Evaluation of Hydropower Project* (DL/T 5441 -2010), the estimated input and output prices of financial evaluation generally do not include the value-added tax price. Otherwise, it should be noted.

4.1 Basic Principles, Price System and Evaluation Contents of Financial Evaluation

1. Price system of financial evaluation

There are two categories of factors that affect price variation: factors of relative price variation and factors of absolute price variation. Relative price is the price comparison between commodities, and absolute price is the price level of commodities in terms of monetary units. The variation of absolute price generally shows the variation of general price, i.e., the general rise of all commodity prices caused by currency devaluation (or inflation), or the general decline caused by deflation.

There are three kinds of prices involved in financial analysis: base price, real price and current price.

(1) The base price, also known as fixed price, is the price in terms of the price level of the base year, regardless of the subsequent price changes. If the base price is adopted, the price of each year in the calculation period of the project is the same. Generally, the year in which the evaluation work is carried out is the base year. The base price is the basis for determining the predicted price of various goods involved in the project, and the basis for estimating the construction investment.

(2) Current price is the market price at any time, in terms of the current price level. The current price is based on the base price, and it is calculated based on estimated price rising rates for different goods during the calculation period.

(3) The real price, in terms of the base year price, it reflects the influence of relative price change factors. It can be obtained by deducting the influence of the change of the general price level from the current price. Only when the price rise rate is greater than the average price rise rate, the real price rise rate of the goods will be greater than 0. This means that the price rise of the goods exceeds the total price rise. If the relative price of goods remains unchanged, that is, the real price rising rate is 0, then the real value is equal to the base value. This means that the current price rising rate of all goods is the same, and equal to the total price rising rate.

2. Pricing principle of financial evaluation

Under market economy, the price of goods varies from place to place and from time to time. During the calculation period, the accurate prediction of it is difficult. According to *Construction Projects Economic Evaluation Method and Parameter* (3rd Edition), simplified method can be adopted when the evaluation conclusion is not affected:

(1) For the output during the construction period, as the predicted period is short, the variation of relative price and total price level shall be considered. As the input are different, so it is difficult to predict separately, and the uncertainty may also increase. Therefore, in practice, it is generally calculated in the form of reserve for price rise (or reserve for price difference).

(2) As for the price of inputs and outputs during the operation period, it is difficult to predict the price rise level in the future or the reliability of the prediction results in the

early stage because of long operation period. Therefore, only the price at the beginning of the operation period is generally predicted, and the same price is used in each year of the operation period.

Most financial evaluation of water conservancy projects and hydropower projects in China are based on the above simplified method. In the project investment during the construction period, price rise factors such as reserve fund for price difference are considered, and the prices of each year during the operation period are the same.

According to *Regulation for Economic Evaluation of Water Conservancy Construction Projects* (SL 72 -2013), for price variation factors of financial prices, different measures are suggested: when analyzing the solvency, the forecast price adopted in each year during the calculation period is the price that considers both the relative price variation and the total price level rise based on the total price level in the benchmark year. Generally, only the construction period is considered. If the financial price does not consider the rise of the general price level due to the restrictions of conditions, sensitivity analysis shall be conducted on the impact of the change of this factor on the project's solvency, i. e., the current price mentioned above. In the analysis of profitability, the forecast price adopted in each year during the calculation period is the price that only considers the relative price variation but not the rising factor of the total price level based on the total price level in the base year, i. e., the real price mentioned above.

In financial evaluation of foreign water conservancy and hydropower projects, in addition to the price rise during the construction period, the rise of project operation cost in the operation period due to inflation should also be considered. Correspondingly, the estimated on-grid price and water supply price also increase year by year.

4.1.3 Financial Evaluation Contents

According to *Construction Projects Economic Evaluation Method and Parameter* (3rd Edition) for operational projects, financial analysis should be based on the preparation of financial analysis statements, calculation of financial indicators, analysis of the financial viability, solvency and profitability of the project, judgment of the financial acceptability of the project, determination of the value contribution of the project to the financial subject and investors, and provision of basis for project decision-making. For non-operating projects, financial analysis mainly analyzes the financial viability of the project.

According to *Regulation for Economic Evaluation of Water Conservancy Construction Projects* (SL 72 -2013), the financial evaluation of water conservancy construction projects includes the analysis of financial viability, solvency, and profitability. The financial evaluations are different for different projects and financial income and costs.

(1) For the project with annual financial revenue greater than annual total cost, the maximum loan capacity should be estimated. Besides, perform a comprehensive financial

evaluation, such as financial viability analysis, solvency analysis and profitability analysis, to judge the profect financial feasibility.

(2) For projects without financial revenue or the annual financial revenue is less than the annual operation cost, financial viability analysis should be made. Besides, policies and measures need to be carried out to maintain project normal operation.

(3) If the annual financial revenue is greater than the annual operation costs but less than the annual total cost, focus on the viability analysis. And conduct solvency analysis according to the specific situation.

Comprehensive utilization water conservancy project, water conservancy project composed of multiple single projects, water supply project with multiple water sources or multiple water receiving areas shall be evaluated. If necessary, the cost, benefit and financial evaluation indicator of each main function, single project and different receiving areas can be calculated, respectively.

Financial analysis can be divided into pre-financing analysis and post-financing analysis. Generally, the pre-financing analysis should be first carried out. After getting reasonable pre-financing analysis conclusions, continue to the post-financing analysis.

For a single-purpose power generation project, calculate the financial income and analyze the financial feasibility of the power station according to the possible on-grid energy and local on-grid price. Otherwise, estimate the on-grid price according to the possible on-grid energy and the target income requirements of the project owner.

For foreign investment projects, financing environment investigation are furtherly needed, including laws and regulations, economic environment, financing channels, tax conditions and investment policies.

4.2 Total Investment Estimation

According to *Construction Projects Economic Evaluation Method and Parameter (3rd Edition)*, the total project investment is the sum of the construction investment, the interest during the construction period and all the working capital, i.e., the investment during construction and operation of the project.

According to *Specification on Economic Evaluation of Hydropower Project (DL/T 5441-2010)*, in hydropower construction projects, the total project investment is the sum of construction investment and interest during construction as well as working capital.

According to *Regulation for Economic Evaluation of Water Conservancy Construction Projects (SL 72-2013)*, the total project investment is the sum of fixed-asset investment and interest during construction period. Comparatively, the total investment approved by the Ministry of Water Resources of the people's Republic of China and the National Development and Reform Commission only includes fixed-asset investment and interest during construction but excluding working capital. This is the difference between water conservancy projects and hydropower projects.

4.2.1 Construction Investment (Fixed Asset Investment)

Construction investment, known as the fixed-asset investment, is the investment which can be directly provided by budget estimate. It is the basic data for project financial evaluation.

The fixed-asset investment of water conservancy project includes the project investment, immigrant investment, and environment investment. The project investment consists of construction investment, purchase and installation of electromechanical & metallic equipment, temporary project investment, independent cost, basic reserve fund and price-difference reserve funds. The environment and immigrant investment consists of land requisition compensation investment, soil and water conservation investment and environmental protection investment. See Figure 4.1 for details.

Figure 4.1 Components of fixed asset investment in water conservancy project

The construction investment of hydropower construction project consists of four parts: the investment of pivotal project, the compensation for land acquisition and resettlement, the independent cost and the reserve cost. The investment in pivotal project includes the investment in auxiliary construction project, construction project, environmental protection and water and soil conservation project, mechanical and electrical equipment and installation project, metal structure equipment and installation project. See Figure 4.2 for details.

4.2.2 Interest During Construction Period

Interest during the construction period is the interest of debt fund. It allows to be included in fixed assets after being put into production. That is, the capitalized interest. Interest during the construction period includes interest on bank loans and other debt funds, as well as other financing costs. Other financing costs are handling charges, commitment fees, management fees and credit incurred in some debt financing loan insur-

4.2 Total Investment Estimation

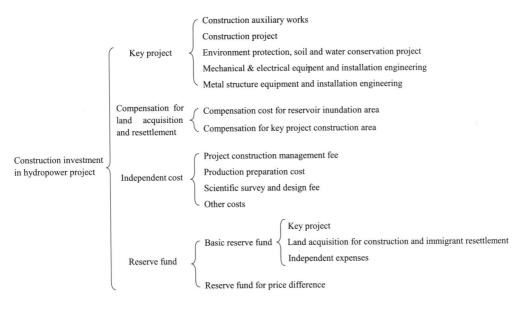

Figure 4.2 Investment composition of hydropower project construction

ance premium, etc. Sometimes, Chinese investors invest in overseas projects. In order to avoid investment losses caused by political turmoil or changes in government laws and regulations in the invested country, the Chinese investors apply for overseas investment insurance premiums to insurance companies. The investment insurance premiums should also be included in the interest during the construction period.

1. Interest during construction period

According to the different interest payment conditions of the creditor, the interest during the construction period is generally divided into three types:

Firstly, there is no need to pay interest before putting into production, but the outstanding interest should be calculated as the principal (i.e. compound interest). Most domestic banks use this type of interest payment.

According to *Regulation for Economic Evaluation of Water Conservancy Construction Projects* (SL 72 - 2013), in order to simplify the calculation of interest during the construction period, it is assumed that the loan will be used in the middle of the year, and the interest will be calculated in half a year, and the interest in each subsequent year will be calculated in the whole year. The interest during the construction period can be calculated according to the following formula:

annual accrued interest = (*accumulated principal and interest of loan at the beginning of the year* + *loan amount of the year*/2) × *annual interest rate*

Secondly, the interest during the construction period must be paid as may be paid. The interest during the construction period can only be paid through the project capital. Some foreign banks adopt this type of interest payment. The calculation method of interest is:

annual accrued interest = (*accumulated principal of loan at the beginning of the year* + *loan amount of they ear*/2) × *annual interest rate*

Thirdly, not only the interest is paid, but the principal is lent by way of interest deduction.

annual accrued interest = (*accumulated principal of loan at the beginning of the year* + *loan amount of the year*/2) × *annual interest rate* / (1 − *annual interest rate*)

The interest during the construction period shall be calculated respectively based on the capital source and financing plan of the project and the production time of various hydraulic structures and equipment. There are different regulations on whether the interest incurred in the early stage of operation period is included in the interest during the construction period.

According to *Specification on Economic Evaluation of Hydropower Project* (DL/T 5441 -2010), all the loan interest before the first unit put into operation is included in the interest during the construction period. As more units are put into operation, the corresponding interest is included in the production cost. The interest division between the construction period and the operation period is based on the production capacity or power generation.

According to *Regulation for Economic Evaluation of Water Conservancy Construction Projects* (SL 72 - 2013), the loan is considered as non - repayment of principal and interest during the construction period, and the interest is calculated on annual basis. The interest during the construction period is calculated by compound interest until the end of the construction period and is included in the total investment.

The annual interest rate of the loan is the loan interest rate published by banks for the same period (if the annual interest is calculated many times, the effective annual interest rate shall be considered). The loan term is considered as the number of repayment years, including grace period and repayment period.

The loan interest rate by banks is generally the annual interest rate. If the interest is calculated by half a year, quarter or month, the nominal interest rate needs to be converted into the effective annual interest rate. The calculation formula is:

$$i = \left(1 + \frac{r}{m}\right)^m - 1$$

Where:

i — effective annual interest rate, i. e. annual interest rate;

r — nominal annual interest rate;

m — the number of annual interest calculation, twice for half a year and four for quarter.

For domestic bank dollar loan, the loan interest rate is composed of LIBOR interest rate and margin. LIBOR is the abbreviation of London Inter-bank Offered Rate. It is the most important benchmark interest rate in the international financial market at present, calculated and published by the British bankers' Association (BBA). LIBOR interest rate is the European dollar capital among London international banks. The interest rate of gold dismantling varies from overnight to 5 years. The most common period is 3 months or 6 months. Each trading day has a different LIBOR value. When calculating loans, banks require LIBOR rate to be the average value of international LIBOR in recent ten years. Many financial companies determine the loan interest rate based on LIBOR value and a certain margin.

2. Service charge

Service charge is the fee that needs to be paid to the creditor's institution in financing. It is generally based on the total amount of debt funds. The service charge rate is determined by banks, generally 1.0% – 1.5%. Different banks have different requirements. The service charge is paid once in the first year of the loan.

3. Commitment fee

Originally, it is the loan borrowed from the world bank. For the effective but unused part, the borrower shall pay the commitment fee, i. e. , the fee for the lender's commitment to provide funds.

At present, domestic banks charge commitment fees to entities that invest abroad. Banks issue a letter of commitment to customers, promising to provide loans within a certain amount. If customers do not use the funds during the loan process, in order to make up for the loss of the bank's inability to use the funds at will due to the commitment of loans, the fees charged by banks to customers are commitment fees. Generally, the committed rate is 0.125% – 1%. The commitment fee shall be calculated based on the unused amount and the actual unused days. It shall be paid quarterly, semi-annually or annually.

4. The management fees

The management fee is a one-time charge for bank loans. It is calculated according to a certain proportion of the total loan. Generally, the rate is 0.1% – 1%. The management fee can be paid by the form of one-time payment when the loan agreement is signed.

5. Overseas investment insurance premium

Overseas investment insurance is an insurance to cover the investment loss caused by political situation of the invested country or the change of government laws and regulations. The overseas investment insurance is to protect the overseas investment of the investor from the loss caused by expropriation, exchange restriction, war and civil disturbance. It is also called political risk insurance or overseas investment guarantee insurance. As an important means of international investment protection, overseas investment insurance is a risk prevention and economic compensation measure taken by the government of an investing country to encourage its capital to invest overseas to obtain high profits. Investors can get compensation from the insurer once suffering investment loss due to political risk in the invested country. Therefore, the overseas investment insurance has similar attributes with the export credit insurance, but it is limited to the export of capital and restricted by the international investment conventions signed by the invested governments with other countries.

At present, overseas investment insurance has been recognized as an effective system to promote and protect international investment. The way of setting up risk fund to manage risk has lower financial pressure than the traditional way of subsidy. And it is also the way of supporting domestic enterprises' overseas investment approved by WTO.

Overseas investment insurance can play an important role in enlarging overseas investment. Many overseas investment projects, especially those in emerging markets and countries in transition economy, have high political risks, and the losses caused by political risks are also huge, so investors often give up their investment plans. Overseas investment insurance can reduce investors' expectation of risk to a certain extent.

Overseas investment insurance is a kind of guarantee insurance provided by the government. It is a kind of "national guarantee" for overseas investors. It is not only implemented or entrusted by the national ad – hoc institutions, but also aimed at the national risk derived from the state power. It is usually not covered by commercial insurance. Overseas investment insurance is a kind of non – profit policy insurance. The insurance is to encourage enterprises to invest abroad and ensure that overseas investment enterprises avoid all kinds of uncertain losses caused by political risks and credit risks.

The risks covered by overseas investment insurance are expropriation, exchange restriction, war and government default. Expropriation refers to the forcible expropriation, confiscation, nationalization, seizure and other acts taken, approved, authorized or agreed by the government of the country where the investment is located. These behaviors need to last for a period and make investors unable to establish or operate the project enterprise or deprive or hinder investors' rights and interests. War refers to wars, civil wars, terrorist acts and other similar acts in the country where the investment is located. The guarantee under the war situation includes the loss of tangible property of the project enterprise caused by the war and the loss caused by the abnormal operation of the project enterprise caused by the war behavior. Exchange restriction refers to the measures implemented by the government of the country where the investment is located to prevent the investor from converting the local currency into convertible currency and/or remitting it out of the country where the investment is located, or to make the investor have to convert the local currency into convertible currency and/or remitting it out of the country where the investment is located at a price much higher than the market exchange rate. Government default means that the government of the country where the investment is located illegally or unreasonably cancels, violates, fails to perform or refuses to recognize the specific guarantee, guarantee or concession agreement related to the investment issued or signed by it.

Insurance amount: equity investment insurance is the initial investment plus not more than twice the retained earnings.

Insurance period: usually up to 15 years, individual 20 years.

Compensation ratio: 90% in general and 100% in some cases. However, if the investor fails to fulfill his contractual obligations, the insurance institution shall not be liable for compensation.

However, if the investor fails to perform his contractual obligations, the insurance institution shall not be liable for compensation. The premium for overseas investment is based on the equity capital of the investor, and the rate is generally 1%.

6. Overseas loan insurance premium

Overseas loan insurance premium, also known as export buyer's credit insurance, is the policy insurance that the export credit agency (ECA) provides loan banks with repayment guarantee under the mode of export buyer's credit financing. The risks include political risks and commercial risks, and the compensation ratio is 95%. The basic contract is the export buyer's credit loan agreement. The currency of insurance is the currency of the loan agreement, generally US dollars. The insured of the export buyer's credit is the loan bank, and the policy holder is generally the exporter or the loan bank.

For foreign hydropower projects financed from Chinese banks, there are two ways to pay the credit insurance premium: one is one-time payment in the first year of the loan period. Based on the sum of principal and interest of the loan, the proportion of the insurance premium is about 7%. If it is paid at the beginning of the loan period, the insurance premium will be higher than 7%. The other is to pay in each year of the loan. The annual loan balance is the base, and the proportion of insurance premium is about 1%. Specifically, the calculation of the premium for overseas loans shall be carried out in consultation with China Export & Credit Insurance Corporation.

4.2.3 Working Capital

Working capital is used for turnover for a long time during the operation period, excluding the working capital temporarily needed. The expanded indicator estimation method or itemized detailed estimation method can be used.

1. Expanded indicator estimation method

The expanded indicator estimation method can be estimated by reference to the proportion of the circulating funds of the same kind of built projects in the sales revenue and operation cost, or the amount of the unit output in the circulating funds.

According to the standard for the value of working capital proposed in the *Regulation for Economic Evaluation of Water Conservancy Construction Projects* (SL 72 -2013), in case of lack of data, the water supply and irrigation projects can be estimated as 1.5 times of the monthly operation cost, or as 8%-10% of the annual operation cost. According to project size, the working capital of hydropower project can be estimated as 10 - 15 yuan/kW of installed capacity.

According to *Interim Provisions on financial evaluation of hydropower construction projects (Trial)*, the working capital of hydropower stations are generally estimated as 10 yuan/kw of installed capacity.

2. Itemized detailed estimation method

working capital = working assets − working liabilities

working assets = receivable account + prepayment + inventory + cash

working liabilities = payable account + accounts received in advance

working capital increase in current year = working capital of current year − working capital of last year

4.2.4 Fixed Assets, Intangible Assets and Other Assets

According to *Construction Projects Economic Evaluation Method and Parameter (3rd Edition)*, after the project is put into operation, the construction investment and the interest during the construction period are composed of three parts, i.e., fixed assets, intangible assets and other assets based on the principle of capital preservation. That is, the construction investment and the in during the construction are composed of the original value of fixed assets, intangible assets and other assets.

construction investment (fixed assets investment) + interest during construction

= original value of fixed assets + original value of intangible assets

+ original value of other assets

For water conservancy construction projects, there are few intangible assets and other assets. The cost of land acquisition for immigrants is embodied in the cost of buildings. The investment of fixed assets and the interest during the construction period are the original value of fixed assets. Also, it can be called the value of fixed assets or the original value of fixed assets.

Fixed assets are houses, buildings, machines, machinery, means of transport and other equipment, instruments and tools related to production and operation with a service life of more than one year.

Intangible assets are identifiable non-monetary assets owned or controlled by enterprises without physical form, including patent right, trademark right, copyright, land use right, non-patented technology, goodwill, etc.

Other assets are various costs that cannot be fully included in current year's profits and losses and shall be amortized in subsequent years. Specifically, start-up costs and improvement costs of rented fixed assets are included.

When it is difficult to calculate intangible assets and other assets, the formation rate of fixed assets can be considered as 100%, namely:

$$construction\ investment + interest\ during\ construction$$
$$= original\ value\ of\ fixed\ assets\ (or\ valve\ of\ fixed\ assets)$$

4.3 Total Cost Estimation

The total cost is all costs to produce products or services during the operation period. It is the sum of the operation cost, depreciation cost, amortization cost and financial cost. The total cost can be calculated by production cost plus period cost estimation or production factor estimation.

1. Estimation of Production cost plus period cost

$$total\ cost = production\ cost + period\ cost$$
$$production\ cost = direct\ material\ cost + direct\ fuel\ and\ power\ cost$$
$$+ direct\ wage + other\ direct\ expenditure$$
$$+ manufacturing\ cost$$
$$period\ cost = management\ fee + operation\ cost + financial\ cost$$

2. Production factor estimation

$$total\ cost = purchased\ raw\ materials,\ fuel\ and\ power\ cost + wage\ and\ welfare\ cost +$$
$$depreciation\ cost + amortization\ cost +$$
$$repair\ cost + financial\ cost + other\ costs$$

For water conservancy and hydropower construction projects, the total cost is generally estimated by the method of production factor estimation. According to the project characteristics, the total cost is generally composed of depreciation, amortization, financial expenses, salaries and welfare (employee compensation), repair and management expenses, insurance expenses, reservoir fund and other expenses. Some projects also include fuel and power expenses, raw water cost, water resource fee, etc. See Figure 4.3 for details.

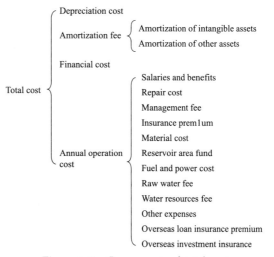

Figure 4.3 Components of total cost

Operation cost, also known as the annual operation cost, is a specific concept in project economic evaluation. It needs to be paid annually to maintain the normal operation of water conservancy construction project. As the main cash outflow during the operation period, it is the part of the total cost excluding depreciation, amortization and financial cost. Specifically, it is composed of salaries and welfare (employee compensation), repair fee, management fee, insurance fee, material fee, reservoir fund, fuel and power fee, raw water fee, water resource fee and other expenses. For overseas investment projects, there are also overseas investment insurance premium and overseas loan insurance premium.

According to the relationship between cost and output, total cost can also be divided into fixed cost and variable cost. Fixed cost, also known as constant cost, is the part of total cost that does not change with the increase or decrease of product output. Variable cost is the part of total cost that changes with the increase or decrease of product output.

4.3.1 Depreciation Cost

The depreciation cost is also known as the depreciation cost of fixed assets. The fixed assets of the project will be worn out over time, and the loss of value is usually compensated by the way of drawing depreciation. According to the provisions of the financial system, the fixed assets of an enterprise shall be depreciated monthly and shall be included in the cost or current profit and loss of the relevant assets. In financial analysis, when the total cost is estimated by the factor of production method, the depreciation of fixed assets can be directly included in the total cost. The depreciation method of fixed assets can be determined by the enterprise itself within the scope permitted by the tax law, generally adopting the straight-line method, including the straight-line method and the workload method. China's tax law also allows the use of rapid depreciation method for certain machinery and equipment. That is, the double declining balance method and the total number of years method.

The depreciation method of fixed assets of water conservancy and hydropower construction projects generally adopts the straight-line method of life average. The annual depreciation cost is equal to the product of the value of fixed assets and the annual depreciation rate.

annual depreciation rate $= (1 - $ *estimated net residual value rate*$)/$*depreciation life*

Due to the large number of water conservancy and hydropower construction projects, the depreciation rate shall be calculated according to the service life of power generation equipment and civil engineering, or the comprehensive depreciation rate shall be adopted. Comprehensive depreciation rate is the average depreciation rate calculated by all fixed assets of an enterprise. The calculation formula is:

$$comprehensive\ depreciation\ rate = \sum_{i=1}^{n} (fixed\ assets\ of\ i^{th}\ classi \cdot annual\ depreciation\ rate\ of\ fixed\ assets\ of\ class\ i)$$

4.3 Total Cost Estimation

/*investment in fixed assets* $\times 100\%$

The comprehensive depreciation rate of hydropower projects is generally 3%–3.3%, and even 4%. The comprehensive depreciation rate of water conservancy projects is generally 2.5%–2%.

See Table 4.1 for depreciation life of main equipment.

Table 4.1   Depreciation Life of Fixed Assets of Industrial Enterprises

Item classification	Depreciable life/a
Ⅰ. General–purpose equipment	
1. Mechanical equipment	10–14
2. Power equipment	11–18
3. Conducting equipment	15–28
4. Transporting equipment	6–12
5. Automation control and instrument and apparatus	
5.1 Automation, semi–automatization control equipment	8–12
5.2 Computer	4–10
General measurement instruments and equipment	7–12
6. Industrial furnace	7–13
7. Equipment and other production tool	9–14
8. Non–production equipment and appliance	
8.1 Tool	18–22
8.2 Television, printer, word processor	5–8
Ⅱ. Special equipment	
9. Special equipment for metallurgical industry	9–15
10. Power industry special equipment	
10.1 Power generation and heating equipment	12–20
10.2 Transmission line	30–35
10.3 Distribution line	14–16
10.4 Power transformation and distribution equipment	18–22
10.5 Nuclear power generation equipment	20–25
11. Special equipment for machinery industry	8–12
12. Special equipment for petroleum industry	8–14
13. Special equipment for chemical and pharmaceutical industry	7–14
14. Special equipment for electronic instrument and telecommunication industry	5–10
15. Special equipment for building materials	6–12
16. Special equipment for textile and light industry	8–14

Continued

Item classification	Depreciable life/a
17. Special equipment for mining, coal and forest industry	7 – 15
18. Special equipment for shipping industry	15 – 22
19. Special equipment for nuclear industry	20 – 25
20. Special equipment for public utility enterprises	
20.1 Tap water	15 – 25
20.2 Gas	16 – 25
Ⅲ. Housing and structures	
21. Housing	
21.1 Production housing	30 – 40
21.2 Corroded production room	20 – 25
21.3 Strong corrosion production room	10 – 15
21.4 Non – production housing	35 – 45
21.5 Cabana	8 – 10
22. Structures	
22.1 Hydropower dams	45 – 55
22.2 Other structures	15 – 25

The depreciation life of water conservancy construction projects is listed in the appendix of *Regulation for Economic Evaluation of Water Conservancy Construction Projects* (*SL 72 -2013*). See Table 4.2 for details.

Table 4.2   Fixed – assets classified depreciation life of water conservancy project

Classification of fixed assets	Depreciable life/a
Ⅰ. Dike, dam, sluice structure	
1. Large – scale concrete, steel reinforced concrete, dike, dam and slice	50
2. Small and medium – sized concrete, steel reinforced concrete, embankment, dam and slice	50
3. Earth, earth – rock mixture local material, embankment and dam	50
4. Concrete, asphalt and other impervious earth, earth rock, rockfill, masonry and other local materials embankment, dam	50
5. Small and medium – sized culvert and sluice	40
6. Semi – permanent gate and dam of wood structure and nylon	10
Ⅱ. Overflow installation	
1. Large concrete and reinforced concrete spillway	50
2. Small and medium – sized concrete and reinforced concrete spillways	40

4.3 Total Cost Estimation

Continued

Classification of fixed assets	Depreciable life/a
3. Concrete and reinforced concrete spillway	30
4. Masonry spillway facilities	20
III. Flood discharge, drainage pipe and tunnel structures	
1. Large concrete reinforced concrete pipe and hole	50
2. Small and medium-sized concrete reinforced concrete pipe and hole	40
3. Unlined pipe and hole	40
4. Masonry pipe and hole	30
5. Brick pipe and hole	20
IV. Water diversion, irrigation and drainage channel and pipe network	
A Large scale	
1. Concrete and reinforced concrete diversion channel	50
2. Water diversion, irrigation and drainage canals of general built earth	50
3. Concrete, asphalt and other protective masonry, impervious canal channel	40
4. Water drop, aqueduct, inverted siphon and other buildings	50
B Small and medium scale	
1. Water diversion, irrigation and drainage channels of general built soil	40
2. Concrete, asphalt and other protective masonry, anti-seepage channel	30
3. Plastic and other non-permanent seepage control channels	25
4. Canal structure, including hydraulic drop, aqueduct, inverted-siphon, regulating sluice, diversion sluice, etc.	30
C Water transmission and drainage network	
1. Ceramic pipe, concrete, asbestos cement pipe network	40
2. Steel pipe and cast-iron pipe network	30
3. Plastic pile, PVC pipe	20
V. Well	
1. Deep well	20
2. Submerged well	15
VI. River regulation and control project	
1. Riprap work and stone revetment	25
2. Spur dike, longitudinal dike and other control and guide works	20
VII. Building structure	
1. Metal and reinforced concrete structures	50
2. Reinforced concrete and masonry composite structure	40
3. Permanent brick and wood structure	30
4. Simple brick wood structure	15

Continued

Classification of fixed assets	Depreciable life/a
5. Temporary civil construction	5
Ⅷ. Metal structure	
1. Steel penstock	50
2. Large gate valve, hoist equipment	30
3. Small and medium-sized gate valve and hoisting equipment	20
Ⅸ. Electromechanical equipment	
1. Large hydraulic turbines	25
2. Small and medium sized water turbine units	20
3. Large scale electric drainage and irrigation equipment	25
4. Small and medium drainage and irrigation equipment	20
5. Small and medium sized mechanical drainage and irrigation equipment	10
Ⅹ. Equipment for power transmission and distribution	
1. Iron tower, steel reinforced concrete pole	40
2. Electric cable, wooden pole line	30
3. Substation equipment	25
4. Power distribution equipment	20
Ⅺ. Water pump and sprinkling equipment	
1. Centrifugal pump	12
2. Deep-well pump	8
3. Submersible pump	10
4. Sprinkler irrigation equipment	6
Ⅻ. Tools and equipment	
1. Production tools, appliances, survey, experiment, observation, research and other instruments and equipment	10
2. Railway transportation equipment, steel water transportation equipment	25
3. Automobile and other mechanical equipment	15
4. Wooden water transport equipment	10

4.3.2 Amortization Expenses

Amortization expenses include the amortization expenses of intangible assets and other assets. According to relevant regulations, intangible assets shall be amortized and included in the cost averagely within the effective service life from the date of use. If the legal term of validity or the beneficial period is stipulated by laws and contracts, the amortization period shall follow its provisions. Otherwise, the amortization period shall meet the requirements of tax law. Generally, the amortization of intangible assets uses the straight-line average method, and residual value is omitted similarly, the amortization of other assets also adopts the straight-line average method, and residual value is omitted. The amortization period shall meet the requirements of tax law.

The amortization expenses of intangible assets are equal to the product of the value of intangible assets and the rate of amortization fee. For other assets, it is the same.

There are few intangible assets and other assets in water conservancy projects and hydropower projects. Generally, intangible assets and other assets are not considered in water conservancy projects and hydropower projects.

4.3.3 Financial Cost

Financial cost are the expenses incurred by the producers and operators to raise funds. Specifically, it consists of long-term loan interest expense, current fund loan interest expense and short-term loan interest expense during the production and operation period, as well as the net exchange loss, service charges of financial institutions and other financial expenses during financing.

1. Long-term loan interest expense

The long-term loan repayment methods include equal repayment of principal and interest, equal repayment of principal and interest on loan, repayment according to ability and repayment according to agreement.

(1) Equal repayment of principal and interest. Equal repayment of principal and interest means that the sum of principal and interest repaid each year during the loan repayment period is equal. The loan interest is decreasing year by year, but the principal repaid is increasing year by year. The annual repayment can be calculated as follows:

$$A = I_c \frac{i(1+i)^n}{(1+i)^n - 1}$$

Where:

A—annual repayment of principal and interest, 10^4 yuan;

I_c—sum of principal and interest of loan at the end of construction period, 10^4 yuan;

i—annual interest rate;

n—the repayment period, calculated from the repayment year.

The Excel-internal function PMT (Rate, Nper, Pv, [Fv], [Type]) can be used for the calculation. In the formula, PMT function syntax has the following parameters:

Rate—interest rate of each period;

Nper—total number of periods (calculation period is year, month, etc., the same as interest rate calculation period);

Pv—the sum that has been recorded at the beginning of the calculation of the loan, or the cumulative sum of the current value of a series of future payments;

Fv—future value, or cash balance that can be obtained after the last payment. If ignored, the value is 0;

Type—optional number 0 or 1, used to specify whether the payment time of each period is at the beginning or the end of the period. At the beginning, Type=0, at the end of the period Type=1. If Type is omitted, the value is assumed to be 0.

Total principal and interest of loan at the beginning of the year=total principal and interest of loan at the beginning of the previous year—repayment of loan principal in the previous year

Annual interest payment=total principal and interest of loan at the beginning of the year×annual interest rate

Annual repayment of principal=annual repayment of principal and interest—annual payment of interest

(2) Payment method of equal amount principal and interest. Equal repayment of principal and interest means that the principal repaid each year during the loan repayment period is equal, the loan interest is decreasing year by year, and the total repayment is also decreasing year by year. The annual repayment can be calculated as follows:

$$A_t = \frac{I_c}{n} + I_c \left(1 - \frac{t-1}{n}\right) i$$

Where:

A_t—repayment of principal and interest in the tth year, 10^4 yuan;

I_c—sum of principal and interest of loan at the end of construction period, 10^4 yuan;

i—annual interest rate;

n—repayment period calculated from the repayment year.

total principal and interest of loan at the beginning of the year=total principal and interest of loan at the beginning of the previous year—repayment of loan principal in the previous year

annual repayment of principal=sum of principal and interest of loan at the end of construction period I_c/repayment period from repayment year n

annual interest payment=total principal and interest of loan at the beginning of the year×annual interest rate

That is, the interest paid in t^{th} year can be calculated as $I_c \left(1 - \frac{t-1}{n}\right) i$.

The annual repayment of principal is $\frac{I_c}{n}$.

(3) Repayment by capacity. Repayment by capacity means that all undistributed profits and depreciation and amortization expenses available for loan repayment are used to repay the loan principal. Repayment interest is equal to principal repayment. The method of repayment according to capacity is used in the calculation of the on-grid price. This method meets the requirements of the loan repayment period in the 1990s, but it is seldom

used at present.

(4) Repayment by agreement. Repayment by agreement, as the earliest repayment method, comes from the World Bank loan project. It is considered that at the early stage of project operation, the production is not fully achieved, and the benefits can not be fully exerted. When the equal amount repayment method or equal principal repayment method is adopted, the early repayment pressure is too large. Therefore, according to the agreement requirements of both parties, the principal and interest shall be paid every year.

2. Interest expense of working capital loan

Working capital loans are repayable at the end of the period and re borrowed at the beginning of the period, and interest is calculated at one-year interest rate. The calculation formula is:

Annual working capital loan interest=current capital loan balance at the beginning of the year×annual interest rate of current capital loan

3. Interest expense of short-term loan

The short-term loan in the financial evaluation refers to the short-term loan incurred due to the temporary needs of funds during the operation period. The calculation of interest is the same as that of the loan interest of working capital. The repayment of short-term loan shall be based on the principle of loan with repayment. The loan in the current year shall be repaid in the next year as far as possible. The short-term loan shall not exceed 5 years according to *Regulation for Economic Evaluation of Water Conservancy Construction Projects* (SL 72 -2013).

4.3.4 Salary and Welfare (Employee Compensation)

Salaries and welfare expenses, known as employee salaries in water conservancy projects, refer to various forms of remuneration and other related expenditures given to obtain services provided by employees.

The composition and value method of employee compensation is introduced in *Regulation for Economic Evaluation of Water Conservancy Construction Projects* (SL 72 - 2013). The employee compensation is composed of employee salary (including salaries, bonuses, allowances, subsidies and other monetary remuneration), employee welfare, labor union funds, employee education funds, housing fund, medical insurance, endowment insurance, unemployment insurance, work-related injury insurance, and students social basic insurance premium such as education insurance premium. Generally, it is calculated as the product of the operation and management staffing and the per capita wage level. The water conservancy project staffing shall be calculated according to the personnel determined by project management. The personnel wage, bonus, allowance, and subsidy are calculated by 1.0 - 1.2 times of the average working capital level of the independent accounting industrial enterprise (state-owned economy) provided by the local statistical department. The staff welfare, labor union funds, staff education funds and housing are calculated. The accrual basis of provident fund and social basic insurance premium shall be determined according to the approved corresponding wage standard. In case of lack of information, Table 4.3 is suggested as a reference for calculation.

Table 4.3   Calculation of salaries and benefits (employee compensation)

Serial number	Item	Rate/%	Calculation base	Note
1	Welfare	14	Total salaries	
2	Union funds	2	Total salaries	
3	Personnel education	2.5	Total salaries	
4	Old-age pension	20	Total salaries	
5	Medical insurance premium	9	Total salaries	
6	Industrial and commercial insurance	1.5	Total salaries	
7	Maternity insurance	1	Total salaries	
8	Unemployment insurance fund	2	Total salaries	
9	Housing provident fund	10	Total salaries	
	Total	62		

4.3.5 Repair Cost

Repair cost is the cost of necessary repair to maintain the normal operation and use of fixed assets and give full play to the use efficiency. According to the size of repair scope and repair interval, it can be divided into major repair and minor repair. Repair cost is included in the total cost. If the amount of repair costs incurred in the current period is large, the method of accrual or amortization can be implemented.

The repair cost can be calculated according to the actual statistics of similar projects in recent three years. For hydropower projects, in the *Interim Provisions on Financial Evaluation of Hydropower Construction Projects (for Trial Implementation)*, for projects without actual data, 1% of the value of fixed assets (i.e., the original value of fixed assets) can be used for estimation. According to *Specification on Economic Evaluation of Hydropower Project (Draft for Comment)*, for projects without actual data, 0.5% of the value of fixed assets (i.e. the original value of fixed assets) can be used for estimation. For water conservancy projects, it is generally estimated as 1% of the value of fixed assets (deducting compensation for land acquisition of immigrants). The value of fixed assets can be calculated according to the results in Table 4.11 and the formula in *Section 4.2.4*. The value of fixed assets (deducting the compensation for land acquisition of immigrants) can be calculated according to the results in Table 4.11 and the formula in

Section 4.2.4, and the compensation for land acquisition of immigrants needs to be deducted. For water conservancy projects without data, values can be taken according to Table 4.6 to Table 4.8.

4.3.6 Management Fee

Project management fee is mainly composed of travel expenses, office expenses, consulting fees, audit fees, litigation fees, pollution discharge fees, greening fees, business entertainment fees, and bad debt losses, etc. It can be calculated according to statistic data of similar water conservancy construction projects in recent three years. For water conservancy projects without data, refer to Table 4.6 – Table 4.8.

4.3.7 Insurance Premium

Insurance premium includes fixed assets insurance and other insurance. Water conservancy projects with operation income can be considered when conditions permit, and the premium shall be determined according to the agreement with the insurance company. For hydropower projects without actual data, according to *Specification on Economic Evaluation of Hydropower Project (Draft for Comments)*, 0.5% of fixed assets (i.e. the original value of fixed assets) can be used for estimation. For water conservancy projects without data, refer to Table 4.6 to Table 4.8. The fixed asset or fixed asset of the calculation base (deducting compensation for land acquisition of immigrants) shall be calculated the same as the repair cost.

4.3.8 Material Cost

Material cost includes the cost of raw materials, raw water, auxiliary materials, spare parts, etc. during the operation and maintenance of water conservancy and hydropower projects. The material cost can be calculated according to the actual statistical data of similar projects in recent three years. For hydropower projects without data, please refer to Table 4.4 and Table 4.5. For water conservancy projects without data, refer to Table 4.6 –Table 4.8. The calculation base value of fixed assets (deducting compensation for land acquisition by immigrants) is the same as the repair cost.

4.3.9 Reservoir Fund

The reservoir fund is the required expenses to support the infrastructure construction and economic development planning in the reservoir area and the resettlement area. Besides, this fund also supports the reservoir area protection project and the production and maintenance of living facilities, and solves other remaining problems of the reservoir immigrants after the reservoir impoundment.

According to *"Interim Measures for the Use and Management of the Fund in the Res-*

ervoir Area of a Large and Medium Reservoir", formulated by the Ministry of Finance, the state integrates the funds for the maintenance of the original reservoir area, the funds for the later stage support of the original reservoir area and the funds for the later stage support of the resettlement undertaken by the operational large and medium – sized reservoirs (referring to the reservoirs and hydropower stations with an installed capacity of 25000 kW and above and generating revenue), and establishes the funds for large and medium – sized reservoir area (referred to as the "reservoir area fund"), which is mainly used to support the implementation of infrastructure construction and economic development planning in the reservoir area and resettlement area, the reservoir area protection project and the maintenance of production and living facilities of the immigrants, and to solve other remaining problems of the reservoir immigrants. The reservoir area fund is raised from the power generation income of large and medium – sized reservoirs with power generation income. According to the actual on – grid electricity sales of the reservoir, the amount of electricity is concerned at the standard of no more than 0.008 yuan/(kW · h).

4.3.10　Fuel and Power Cost

The cost of fuel and power mainly includes the cost of pumping electricity during the operation of water conservancy projects, the cost of heating in winter in the northern region and other fuel needed. The electricity charge for pumping shall be calculated and determined according to the characteristics of the pump station, the amount of water pumped and the price of electricity. The heating cost shall be calculated based on the heating building area. The cost can be analyzed and calculated according to the statistical data of similar water conservancy construction projects in the near three years. For water conservancy projects without data, values can refer to Table 4.6 to Table 4.8. The fixed asset value of the calculation base (deducting the compensation for land acquisition by immigrants) is calculated the same as the repair cost.

4.3.11　Raw Water Charges

Raw water charges are the water fee to be paid for taking water from other water source projects. If the water diversion project takes water from a reservoir not built for the project, it needs to pay the raw water fee. If the water is taken directly from a river or lake, or the water intake facilities and the water transmission project are integrated, it does not need to pay the raw water fee. The raw water fee is the product of the raw water price and the water intake amount. The raw water price can be verified by cost plus profit.

4.3.12　Water Resources Fee

According to *Water Law of the People's Republic of China*, units and individuals that directly take water from rivers, lakes or underground shall, in accordance with the provisions of the national water taking license system and the system of paid use of water resources, apply to the water administrative department or the river basin administrative organ for a water taking license, pay the water resources fee and obtain the water taking right. However, small amount of water for domestic life, sporadic raising, captive breeding of livestock and poultry for drinking, etc. shall be excluded.

The standard of water resources fee is determined by the water administrative department at or above the county level in the play where the water intake is located. Generally, it varies according to the type of water supply source and water industry, and the more water resources are scarce, the higher the water resource collection standard is. Water resource fee of water supply project is the product of water intake and water resource fee collection standard, while that of conventional hydropower station is the product of annual power generation and water resource fee collection standard.

On July 6, 2009, the National Development and Reform Commission, the Ministry of Finance and the Ministry of Water Resources jointly issued the water resources fee collection standards for water conservancy projects. It came into effect on September 1, 2009. For water conservancy projects directly under the central government and across provinces, autonomous regions and municipalities directly under the central government for water intake, the water resources fee collection standards shall be examined and approved by the river basin administration and shall be formulated and adjusted by the National Development and Reform Commission together with the Ministry of Finance and the Ministry of Water Resources. For other water conservancy projects, the standards shall be formulated and adjusted by the competent price departments of provinces, autonomous regions and municipalities directly under the central government together with the competent financial and water administrative sectors at the same level, submitted to the people's governments at the same level for approval, and submitted to the National Development and Reform Commission, the Ministry of Finance and the Ministry of Water Resources for the record.

The water resources fee collection standards for water conservancy projects directly under the central government or across provinces, autonomous regions, and municipalities directly under the central government approved by the river basin management agency are: (1) water resources fee for agricultural production is temporarily exempted. (2) Non-agricultural water supply (excluding water supply for power generation) shall be temporarily implemented in accordance with the current standards of the place where the water intake is located. (3) Water resources fee for hydropower generation is 0.003 - 0.008 yuan/ (kW·h) cents per. If water resources fee for hydropower generation at the province, autonomous regions, and municipalities where the water intake is located is lower than 0.003 yuan/(kW·h), it shall be 0.003 yuan. If it is higher than 0.008 yuan, it shall be 0.008

yuan. If it is between 0.003 - 0.008 yuan, it shall remain unchanged.

The water for power generation by pumped storage is exempted from water resource fee temporarily.

4.3.13 Other Expenses

Other expenses are directly related to the production and operation of water supply and power generation in the operation and maintenance of water conservancy and hydropower projects. They are composed of project observation expenses, water quality detection expenses, temporary facilities expenses, etc., in addition to staff salaries and material expenses. The value can be calculated according to the recent investigation data of similar projects. For hydropower projects without data, please refer to Table 4.4 and Table 4.5. For water conservancy projects without data, values can refer to Table 4.6 to Table 4.8. The calculation base value of fixed assets (deducting compensation for land acquisition by immigrants) is the same as the repair cost.

4.3.14 Overseas Investment Insurance Premium

The overseas investment insurance premium is based on the investors' equity of Generally, the rate is 1%. See *Section 4.2.2* for details.

4.3.15 Overseas Loan Premium

At present, the overseas loan insurance premium is generally paid in one time in the first year of the loan period (or the construction period is paid by years). If the operation period also needs to be paid, it is generally based on the loan balance, and the payment proportion is generally 1%. The specific calculation of the premium for overseas loans shall be carried out in consultation with China export and credit insurance corporation. See *Section 4.2.2* for details.

4.3.16 Reference for Parameter Value of total Project Cost

In 1994, according to *Interim Provisions on Financial Evaluation of Hydropower Construction Projects* (for Trial Implementation) when there is no data, see Table 4.4 for the value standard of material cost and other costs.

Table 4.4   Quota of material cost and other costs

Installed capacity/MW	Material cost/(yuan/kW)	Other cost/(yuan/kW)
≤250	5	24.0
251 - 500	3.1	15.6
501 - 750	2.2	6.4

4.3 Total Cost Estimation

Continued

Installed capacity/MW	Material cost/(yuan/kW)	Other cost/(yuan/kW)
751 – 1000	1.7	4.4
1001 – 3000	1.4	2.4
≥3001	1.1	2.0

According to *Specification on Economic Evaluation of Hydropower Project* (DL/T 5441-2010), there is no value standard for material cost and other costs. According to *Specification on Economic Evaluation of Hydropower Project* (Draft for Comments) (2008), when there is no data, the value standard for material cost and other costs is shown in Table 4.5.

Table 4.5　　　　　　　　Quota of material cost and other costs

Installed capacity/MW	Material cost/(yuan/MW)	Other costs/(yuan/MW)
≤300	4000	35000
300 – 1200	3500	30000
1200 – 3000	3000	25000
>3000	2500	20000

The calculation rates of reservoir projects, dike engineering, water supply and irrigation projects without data are summarized in *Regulation for Economic Evaluation of Water Conservancy Construction Projects* (SL 72-2013). See Table 4.6 to Table 4.8 for details. For the original value of fixed assets as the calculation base, it is calculated according to the original value of fixed assets after excluding the compensation for land occupation and inundation.

4.3.17 Total Cost and Operation Cost of Special Supporting Power Transmission and Transformation Project

In some hydropower stations, special power transmission and transformation projects are also included in the cost of the power station. Specifically, it is the sum of depreciation cost, interest expense and operation cost. The depreciation cost and operation cost of special supporting power transmission and transformation projects can be calculated as 4% and 3% of the investment in special supporting transmission and transformation projects, according to *Interim Provisions on Financial Evaluation of Hydropower Construction Projects* (Trial Implementation).

Table 4.6 The calculation rates of reservoir projects

Serial number	Cost item	Charges standard	Calculation basis				Note
			Generation revenue	Flood control	Water supply (including irrigation)		
1	Material fee	Power generation 2–5 yuan/kW Flood control and water supply 0.1%	Installed capacity	Original value of fixed assets	Original value of fixed assets		The original value of fixed assets does not include the compensation for land occupation and inundation
2	Fuel and power costs	0.10%	Original value of fixed assets	Original value of fixed assets	Original value of fixed assets		
3	Repair cost	1%	Original value of fixed assets	Original value of fixed assets	Original value of fixed assets		
4	Staff salary	162%	Gross salary	Gross salary			
5	Management fee	1–2 times	Staff salary	Staff salary			
6	Reservoir fund	0.001–0.008 yuan/(kW·h)	On-grid energy				
7	Water resources fee	According to related regulations of different provinces	Annual power generation		Annual water diversion		
8	Other costs	Power generation 8–24 yuan/kW Flood control and water supply, 10%	Installed capacity	Sum of items 1–4			24 yuan/kW, if the installed capacity $<30\times10^4$ kW. Otherwise, 8 yuan/kW
9	Fixed-asset insurance premium	0.05%–0.25%	Original value of fixed assets				If there is an agreement with the insurance company, the agreement shall be implemented
10	Depreciation cost (amortization cost)	Drafted according to depreciation period (amortization period)	Original value of fixed asset, and other assets				

Note:
1. When the cost of each function of multi-purpose reservoir needs to be calculated separately, the original value of fixed assets allocated to each function shall be taken as the calculation basis.
2. On-grid energy of hydropower station = annual generating capacity × (1 − auxiliary power consumption rate) × (1 − transmission and transformation loss rate).
3. The original value of fixed assets in this table is calculated according to the original value of fixed assets after excluding the compensation for land occupation and inundation.

4.3 Total Cost Estimation

Table 4.7 Calculation rate of annual operation cost of dike project

Method	Cost item	Charging standard			Calculation basis
		First level dike	Second level dike	Third or lower dike	
I	Project maintenance cost	$6\times10^4 - 8\times10^4$ yuan/km	$4\times10^4 - 6\times10^4$ yuan/km	$3\times10^4 - 4\times10^4$ yuan/km	Dike (or river) length
	Management fee	8×10^4 yuan/km	6×10^4 yuan/km	5×10^4 yuan/km	Dike (or river) length
II	Project repair cost	1.0%	1.2%	1.4%	Original value of fixed assets and other assets
	Management fee	0.5%	0.4%	0.3%	

Note: 1. Project maintenance cost includes repair cost, material cost, fuel and power cost and other costs related to project repair and maintenance; management cost includes salaries and welfare expenses, management fees, other expenses and other expenses related to project management.

2. When the fixed assets are used as the base for calculating the annual operation cost, the original value of the fixed assets is used for the new dike projects, and the revaluation value of the fixed assets is used for the existing projects.

3. Small-scale culvert gate and other buildings along the embankment project can be regarded as a whole with the dike project, and the cost is calculated according to the relevant rate of the dike project.

Table 4.8 The calculation rates of reservoir projects, dike engineering, water supply and irrigation projects

Serial number	Cost item	Standards of fees				Calculation basis
		Water diversion project			Pumping station project	
		Pipe culvert	Channel	Tunnel		
1	Project maintenance cost	1.0%-2.5%	1.0%-1.5%	1.0%	1.5%-2.0%	Original value of fixed assets
2	Management fee	1.0%	0.5%	0.3%	1.0%	Original value of fixed assets
3	Water pumping fee				Electricity price	Pumping capacity, delivery head
4	Water resources fee	Water resources fee price yuan/m³				Annual average water diversion
5	Raw water fee	Raw water price yuan/m³				Purchase of raw water
6	Fixed assets insurance premium	0.05%-0.25%				Original value of fixed assets
7	Depreciation cost	3%-4%	2%-2.5%	2%	3%-3.5%	Original value of fixed assets

Note: 1. Project maintenance cost includes repair cost, material cost, fuel and power cost and other costs related to project repair and maintenance. The management fee includes staff salary, management fee and other expenses related to project management, which is calculated according to the proportion of fixed assets.

2. The buildings and small-scale pumping stations along the water transmission trunk line can be regarded as a whole with the water conveyance project, and the cost is calculated according to the relevant rates of the water conveyance project.

3. The water resources fee shall be calculated according to the water quantity of the water diversion canal head section of the water supply or irrigation project, and other intermediate fees shall not be calculated again. The price of water resources fee shall be implemented according to the relevant provisions of the water administrative departments of various provinces and cities.

4.4 Water Supply Price and On–grid Price

4.4.1 Water Supply Price

1. Principles of water supply price of water conservancy project

Under the socialist market economy system, the water price forming mechanism system needs to meet the requirements of the law of value. It should be conducive to rational development and use of water resources, the optimal allocation of water resources, sustainable use of water resources, the prevention and control of water pollution. It should help promote water conservation, improve the efficiency of water use as well as the ecosystem, and support sustainable development of economy and society. At the same time, specific conditions of places and consumer groups, and basic living needs, and affordability of low–income groups need to be considered, to ensure basic domestic water.

According to *Water Supply Price Management Measures for Water Conservancy Projects*, the main points of the water price verification principles and methods are as follows:

(1) Water supply price of water conservancy project consists of water supply production cost, expense, profit, and tax.

Water supply production cost is the direct salaries, direct material cost, other direct expenses, depreciation cost of fixed assets, repair cost, water resource fee and other manufacturing expenses in normal water supply production. Water supply production expense is the reasonable selling expense, management fee and financial expense incurred for organizing and managing water supply production and operation. Profit is the normal supply by water supply operators. The reasonable income from water production and operation shall be determined based on the profit rate of net assets. Tax, included in water price, is the part that the water supply operator shall pay according to national tax law.

(2) The water supply price of water conservancy project adopts unified policy and hierarchical management. Government–guided price or government pricing is implemented according to different situations. For private water conservancy projects encouraged by the government, the government–guided price shall be applied. For other water conservancy projects, the government fixed price shall be applied.

(3) The water supply price of water conservancy projects shall be set according to the principles of compensation cost, reasonable income, high quality, good price, and fair burden. It shall be adjusted in time according to the change of water supply cost, expenses and market supply and demand.

(4) For water conservancy projects with similar engineering conditions, geographical

environment and water resource conditions in the same water supply area, the water supply price shall be determined according to the region. The specific scope of the water supply area shall be determined by the competent department of water administration at the provincial level and the competent department of commercial price. The price of water supply for other water conservancy projects shall be determined according to individual projects.

(5) The assets, costs and expenses of water conservancy projects shall be reasonably apportioned and compensated by category in various purposes such as water supply, power generation and flood control. The costs and expenses allocated for water supply of water conservancy projects shall be compensated by the water supply price.

(6) For water conservancy and water supply projects constructed with loans and bonds, the price of water supply shall enable the water supply operators to have the ability to compensate costs, expenses, repay loans and bonds' principal and interest and obtain reasonable profits during the operation.

(7) According to the national economic policy and the bearing capacity of water users, the water supply for water conservancy projects is priced by classification. The water supply price of water conservancy project is divided into agricultural water price and non-agricultural water price according to the object of water supply. The price of agricultural water shall be determined according to the principle of compensating the production cost and expense of water supply, excluding profit and tax. The price of non-agricultural water is based on the compensation of water supply production cost, expenses and tax. The profit is calculated and drawn according to the net assets of water supply. The profit margin is determined by the long-term loan interest rate of domestic commercial banks plus 2%-3% of that.

(8) The water supply of water conservancy project should gradually carry out the two-part water price system which combines the basic water price and the metered water price. The basic water price shall be determined according to the principle of compensating the direct wage of water supply, management fee and 50% depreciation cost and repair cost. The measured water price shall be determined according to the principle of compensating other costs and expenses, such as water resource fee, material fee, etc. other than the basic water price, and included in the specified profits and taxes.

2. The method of water price determination for water conservancy project

In financial evaluation, clear requirements for water supply price of water conservancy projects are put forward according to *Regulation for Economic Evaluation of Water Conservancy Construction Projects* (SL 72 - 2013). The water price shall be analyzed and proposed according to market price of similar products, project cost price, profit price, user affordable price, approved price approved by competent department, price agreed by both parties, etc. At least, five alternative methods can be adopted for the formulation of water price schemes:

(1) Refer to the water supply price of similar water conservancy construction projects constructed soon and perform the prediction according to the development level and planning of regional national economy, as well as the development and utilization of water resources.

(2) The water supply cost shall be calculated according to the project cost sharing results. It shall be determined according to the cost and investment profit requirements.

(3) Consider the user's willingness to pay and the ability to pay.

(4) Both parties agree on the price of water conservancy products.

(5) The policy price approved by the competent price department or relevant government departments.

In the design of water supply price, cost water price, current water supply price of water users, water price forecast based on the rising trend of water price, water price with little profit for the project, profit water price and affordable water price should be considered. In the measures for the administration of water supply price of water conservancy projects, the profit rate and water price of water supply projects are determined by adding 2 to 3 percentage points to the long – term loan interest rate of domestic commercial banks. The financing scheme has not been determined at the early stage, and the profit rate of net assets water price is the water price of a specific year after project completion. Therefore, the profit rate of net assets water price is not mentioned here in this book. Only the profit water price with dynamic analysis is proposed.

1) Cost water price. Specifically, it includes operation cost water price and total cost water price. Operation cost water price is the quotient of annual operation cost and water supply quantity of the project. Total cost water price is the quotient of total cost and water supply quantity of the project.

2) Current water price. It mainly analyzes the current water price of tap water and water supply price of water conservancy projects in the receiving area and nearby areas of the project.

3) Low profit water price. According to the management method of water supply price of water conservancy project, the water supply price of water conservancy project consists of water supply production cost, expense, profit, and tax. Profit is the reasonable income obtained by water supply operators from normal water supply production and operation. It is determined according to the profit rate of net assets. According to different project financing schemes and loan repayment methods, the low – profit water price is different in each year during the operation period. At the early planning stage, due to the uncertainty of loan proportion, loan period and other financing schemes, it is suggested in this book that the low profit water price should be calculated according to the total investment return rate before financing the water supply project, and the specific calculation

4.4 Water Supply Price and On-grid Price

formula is as follows:

$$low\ profit\ water\ price = (total\ cost\ before\ financing\ of\ water\ supply\ project + fixed\ asset\ investment\ of\ water\ supply\ project \times total\ return\ on\ investment/designed\ water\ supply) / [(1 - value\ added\ tax\ rate) \times (1 + sales\ tax\ additional\ tax\ rate)]$$

4) Profit water price. The profit water price is mainly the water price based on FIRR of capital fund during the operation period according to water supply income, the investment apportioned according to water supply function and financing proportion.

5) Affordable water price. There is no clear boundary between the water price bearing capacity of different water users. the water price bearing capacity is determined by analyzing the proportion of water expenditure occupying the disposable income or the proportion of total income.

The affordable water price should be based on the social and economic investigation and water use investigation of the water use area, the data of the current year, the analysis and prediction results, and the relevant research results or experience coefficients at home and abroad. It aims to estimate the affordability of agricultural, industrial, and urban residents' water users to the water prices.

a. Analysis standard of water price bearing capacity of domestic water for residents.

According to the research results of World Bank and some international loan institutions, it is feasible for households or individuals to spend 3%-5% of their income on water. According to the Research Report on Urban Water Shortage issued by the Ministry of Construction in 1995, If the water fee accounts for 1% of the household income, it has little psychological effect on the residents; If the water fee accounts for 2% of the household income, it has certain impact; If the water fee accounts for 2.5% of the household income, it will cause the attention of the residents for water use and save water; If the water fee accounts for 5% of the household income, it will have a greater impact on the residents' water use, promoting them to save water; If the water fee accounts for 10% of the household income, it will have a large impact on the water consumption of residents, and the reuse of water will be considered. According to domestic experience, water fee generally accounts for 1%-2.5% of the disposable income. The proportion of water fee to disposable income of residents is 1.5%-2%, more appropriate. In northern area, where water resources are scarce, the proportion can be increased to 2.5%.

According to *Regulation for Economic Evaluation of Water Conservancy Construction Projects* (SL 72 -2013), if the annual water resources fee of urban residents' accounts for 1.5%-3% of their annual disposable income, it can be accepted by water users; if the proportion of water resources fee exceeds this range, the market price leverage will be enhanced, and users will take further measures to strengthen water conservation.

b. Standard for analysis of water price bearing capacity of industrial water.

The affordability of industrial enterprises to water price is mainly analyzed according to the proportion of industrial water cost to industrial output value. According to the research results of the World Bank and some international loan institutions, the proportion of water fee of urban industrial water in the total output value of urban industry is about 0.6%. According to the data of profit rate of output value of industrial enterprises in China Statistical Yearbook 2000, the cost of industrial water can be controlled within 1.5% of the industrial output value if other conditions remain unchanged. This can guarantee that the capital profit margin of industry is higher than the bank loan interest rate; If the industrial water fee accounts for 3% of the output value, it will cause enterprises to pay attention to water consumption; If it reaches 6.5%, it will cause enterprises to pay attention to water saving. Enterprises not only save water and use water reasonably, but also take initiatives and measures such as waste water recycling, pollution reduction and efficiency increase.

In water shortage areas, appropriate water expenditure can promote enterprises to save water and use water reasonably. Also, it helps to take positive measures to improve water use efficiency. For example, in water shortage areas it helps to adjust the product structure and industrial structure according to water resources carrying capacity. The value of industrial water expenditure index should be determined according to water resources situation. In northern China, it is generally 2%-2.5%.

According to *Regulation for Economic Evaluation of Water Conservancy Construction Projects (SL 72 -2013)*, it is accepted by water users when the water fee of urban industrial enterprises accounts for less than 2%-5% of the industrial output value. Otherwise, the market price leverage will increase, and water users will take further measures to strengthen water conservation.

c. Standard for water price bearing capacity of agricultural water.

Due to the impact of water users' bearing capacity, the standards of agricultural water cost in the world are generally low. For example, Indian government stipulates that the irrigation cost in India should not exceed 50% of the net income of farmers. Generally it should be controlled at 5%-12%. For some other countries, the proportion of irrigation water cost in the efficiency of irrigation production increase is taken as the practical and feasible standard of irrigation water cost.

According to the analysis and study on irrigation area in China, results are as follows: the percentage of agricultural water cost to agricultural production cost is 20%-30%. The percentage of water cost to the output value per mu is 5%-15%. The percentage of water cost to the benefit of irrigation increase is 30%-40%. The percentage of water cost to the average net income per mu is 10%-20%.

According to *Handbook of Hydraulic Structure Design* (*Second Edition*), the investigation results of agricultural water price, tax revenue and expenditure of irrigation area and per capita net income of farmers in some grain production areas in central and western China show that water fee often accounts for more than 10% of the average input per mu. A considerable part of agricultural water supply cost is higher than the actual water price. If the water price is calculated according to the principle of compensation cost, most areas exceed the bearing capacity of farmers. According to relevant analysis, when the agricultural water fee accounts for 4%-8% of the annual net income, farmers generally think that the water price is reasonable or basically reasonable, and they are willing to pay the water fee, indicating that the water price is within the range of farmers' bearing capacity. When the proportion of agricultural water fee to the average annual net income exceeds 6%, farmers in the irrigation area will think that the water price is high. The percentage of 8% is too high and difficult to accept for most farmers. Therefore, 8% is the upper limit of farmers' ability to bear the agricultural water price.

4.4.2 On-grid Price

The on-grid price of hydropower station includes capacity price and electricity price. In China, the electricity price is adopted.

Since the 1980s, The principal and interest the main basis of financial evaluation the principal and interest repayment price has been used as in China's hydropower construction. Repayment price has two prices during the operation period. The on-grid meeting the loan repayment requirements is higher during the loan repayment period, and the price calculated according to the investment profit rate after the loan repayment is lower. At the end of the 20th century, according to the opinions of the State Planning Commission, the unit put into operation in the future will be changed to the average on-grid price determined according to the whole operation period.

In areas with rich hydropower resources, the electricity price in wet and dry seasons and the time-sharing electricity price in peak and valley are adopted. For these areas, the comprehensive on-grid price needs to be calculated based on long series of power generation data and the balance results of power generation. The on-grid price can be current on-grid price, or the forecast price based on current price system. Otherwise, the prices variation factor can be considered.

Currently, China's power grids in different provinces have different requirements on the on-grid price. For example, The on-grid price is 0.222 yuan/(kW·h) (including value-added tax). In Xinjiang, rich in coal resources, the on-grid price of hydropower is generally not more than 0.235 yuan/(kW·h) (including value-added tax). In Sichuan Province, the benchmark price is implemented. And the benchmark price is divided into peak-valley price and flood season-dry season price. See Table 4.9 for details.

Table 4.9 Relation between on-grid price and benchmark price in different periods of Sichuan Province

Item	Wet season (Jun.-Oct.)	Normal water period (May, Nov.)	Dry season (Dec.-next Apri.)
Peak hours	1.001	1.335	2.003
Normal time	0.75	1	1.5
Trough period	0.375	0.5	0.75

For hydropower station of comprehensive utilization of water conservancy project, according to *Regulation for Economic Evaluation of Water Conservancy Construction Projects* (SL 72-2013), there are some requirements for the on-grid price in financial evaluation. The on-grid price shall be based on the factors and conditions such as market price of similar products, project cost price, profit price, user affordable price, price approved by the competent department, price agreed by both parties, etc. At least the following six methods shall be adopted in making schemes of on-grid price:

(1) With reference to the on-grid price of similar hydropower construction projects constructed soon, the prediction is made according to the development level and planning of regional national economy, as well as the situation of power supply and demand.

(2) Calculate the cost according to the apportioned expenses. And formulate the on-grid price schemes according to the requirements of cost and investment profit.

(3) Formulate the on-grid price schemes based on user's willingness to pay and the ability to pay.

(4) Both parties agree on the price of water conservancy products

(5) Formulate the on-grid price as the policy price approved by the competent price sector or relevant government sectors.

(6) Estimate the on-grid price, according to the requirement that the return on total investment of hydropower station should be 8%-10%, and the FIRR of capital should meet the owners' requirements.

In many countries, the on-grid price of hydropower stations is basically the same as that of China in 1980s. For example, in Pakistan and Indonesia, the on-grid price is calculated respectively during and after the loan repayment period. During the loan repayment period, the on-grid price should be estimated considering the project operation and maintenance cost, repayment of principal and interest, and the owners' expected income. After the loan repayment period, the on-grid price should be estimated considering the project operation and maintenance cost as well as the owner expected earnings. Differently, the average on-grid price considering the social discount rate is adopted for the estimation of on-grid price during the operation period. During the study of the project feasibility and the bidding of the owner, the on-grid price is taken as the evaluation standard. See Chapter 11 for the estimation of hydropower on-grid price in Pakistan.

4.5 Financial Evaluation

According to *Construction Projects Economic Evaluation Method and Parameter (3rd Edition)*, financial evaluation mainly analyzes financial viability, solvency, and profitability. For flood control and waterlogging control projects without profitability, such as levees, waterlogging pumping stations and other projects, the financial viability of the project is mainly investigated.

4.5.1 Financial Income Calculation

The financial income of water conservancy and hydropower construction project includes sales income and subsidy income during the project implementation. Financial income only calculates direct income, not indirect income. Direct income includes water supply income, power generation income and other income.

1. Water supply income

water supply income = *water supply* × *water price* (*excluding VAT*)

Water supply income includes urban water supply income, irrigation water supply income, rural living water supply income, etc.

For the urban water supply project, because it takes a certain time to reach the production capacity, the influence of the initial water supply capacity to reach the production capacity needs to be considered when calculate the water supply income.

2. Power generation

Power generation income of hydropower station includes capacity income and electricity income. At present in China, electricity revenue is calculated.

(1) Electricity revenue

power generation income = *on - grid energy* × *on grid price* (*excluding VAT*)

on - grid energy = *plant power supply* × (1 − *loss rate of special supporting transmission and transformation lines*)

auxiliary power supply = *effective power* × (1 − *auxiliary power rate*)

effective energy = *design energy output* × *effective energy utilization coefficient*

For designed power generation, long series of multi-year average power generation are generally adopted.

According to *Interim Provisions on Financial Evaluation of Hydropower Construction Projects (Trial)* (1994), when there is no data, the loss rate of special supporting transmission and transformation lines is generally taken as 2%, and the auxiliary power rate is taken as 0.2%.

Effective power generation is the average annual power generation expected to be absorbed by the power grid in design. In the power grid with significant hydropower ratio, due to the limitation of the minimum output of thermal power technology, the hydropower station as the base load in the wet season is forced to abandon the water for peak load regulation. Water is abandoned when the daily load is low. Therefore, special attention should be paid to the value of effective power utilization coefficient to avoid exaggerating or reducing the generation revenue.

(2) Capacity income.

power station capacity income = power station required capacity × capacity price

The capacity price can be determined according to the grid regulations of the power station.

3. Other income

Other income of the project includes tourism revenue, aquaculture income, etc.

(1) Tourism income. Tourism income is equal to the sum of ticket income and the income of various tourism fee service items. Ticket income is the product of the number of tourists and ticket price.

(2) Aquaculture income. If the reservoir supplies water to farmers, the aquaculture income can be calculated according to the calculation method of water supply income. If the reservoir provides breeding space, the breeding income is equal to the product of the provided breeding space and unit space price.

For the water conservancy and hydropower projects with navigation function, even if it provides some convenience for shipping, generally the shipping fees shall not be charged. If the individual project has the right of shipping charge, the shipping income shall be calculated according to the approved charging items and charging standards.

4.5.2 Taxes

For water conservancy and hydropower projects in China, the taxes of include value-added tax, sales tax surcharges, income tax, business tax for some water supply projects, and interest tax and dividend tax for overseas investment projects.

1. Value-added tax

According to *"Provisional Regulations of the People's Republic of China on Value-added Tax"* (implemented on January 1, 2009), *the output tax of electric power products = sales amount × tax rate*. The tax rate of value-added tax for hydropower stations greater than or equal to 50 MW is 17%. The tax rate of value-added tax for hydropower stations less than 50 MW is 6%. The VAT rate of water project is 13%. Value-added tax is non-price tax and only serves as the basis for calculating sales tax surcharges.

2. Business tax

According to *"Notice of the State Planning Commission of the Ministry of Finance on Converting Part of the Administrative Fees into Operational Service Fees (prices)"* (CZ [2001] No. 94), the water charges of water conservancy projects are converted from administrative fees to operational service fees. Therefore, water fees collected by water conservancy engineering units from water users are the income for providing natural water. According to current circulation tax policy, no value-added tax shall be levied, and business tax shall be levied according to the tax items of "service industry". According to the above regulations, the business tax rate is 5%.

3. Sales tax surcharges

Sales tax surcharges include urban maintenance and construction tax, education surcharges and local education surcharges. It belongs to the intra price tax.

According to *Interim Regulations of the People's Republic of China on Urban Maintenance and Construction Tax* (GF [1985] No.19), to strengthen the maintenance and construction of cities, and expand as well as stabilize the sources of urban maintenance and construction funds, the urban maintenance and construction tax is collected. The tax is calculated based on product tax, value-added tax, and business tax, and paid simultaneously with the product tax, value-added tax, and business tax. The urban maintenance and construction tax rate are 7% and 5% for taxpayers living in the urban area and in the county or town, respectively. Otherwise, the tax rate is 1%. There is no product tax for water conservancy and hydropower projects. The urban maintenance and construction tax for water conservancy and hydropower projects is based on value-added tax and business tax.

According to *Interim Provisions on the Collection of Educational Surcharges* (issued by the State Council on April 28, 1986), collect educational surcharges in order to implement *Decision of the Central Committee of the Communist Party of China on the Reform of Education System*, accelerate the development of local education, expand the fund sources for local education funds. The educational surcharges are based on the value-added tax, business tax and consumption tax that are paid by various units and individuals. The additional rate of education fee is 3%. It shall be paid together with value-added tax, business tax and consumption tax. There is no product tax for water conservancy and hydropower projects. The education surcharges are based on value-added tax and business tax.

According to *The Circular of the Ministry of Finance on Issues Related to the Unification of Local Education Additional Policies* (CZ [2010] No.98), unify the local education additional collection standards in order to implement *The Outline of the National Medium and Long Term Education Reform and Development Plan* (2010 - 2020), further standardize and expand the financing channels of financial education funds and support local education development. The collection standard of local education surcharges shall be 2% of the value-added tax, business tax and consumption tax that are paid by units and individuals (including enterprises with foreign investment, foreign enterprises, and foreign individuals). For provinces which have collection standard less than 2% approved by the Ministry of Finance, the collection standard should be adjusted to 2%. There is no product tax for water conservancy and hydropower projects. Also, the local education surcharges are based on value-added tax and business tax.

4. Corporate income tax

"The Enterprise Income Tax Law of the People's Republic of China", which came into effect on January 1,2008, has a tax rate of 25%.

The total income of an enterprise in each tax year shall be the taxable income after deducting the non-taxable income, tax-free income, various deductions, and the losses allowed to be made up in the previous year.

Enterprise income obtained from various sources in monetary and non-monetary forms is the total income. The total income consists of: ①income from sales of goods; ②income from provision of services; ③income from transfer of property; ④income from equity investment such as dividends and bonuses; ⑤interest income; ⑥rental income; ⑦income from royalties; ⑧income from donation; ⑨other income.

The following incomes are non-taxable incomes: ①financial appropriation; ②administrative fees and government funds collected and incorporated into the financial management according to law; ③other non-taxable incomes stipulated by the State Council.

The expenses related to enterprise income are composed of costs, expenses, taxes, losses, and other expenses. These expenses can be deducted during the calculation of taxable income.

Some areas enjoy preferential policies of income tax reduction and exemption at the initial stage of project operation. Details shall be in accordance with local policies at that time.

5. Interest tax

For overseas loan projects, in the process of loan repayment in some countries, the project owner needs to pay interest tax when paying interest. The interest tax is calculated based on the interest on loan repayment, and the interest tax rate in most countries is 10%.

In the calculation of financial evaluation, the interest tax can be placed in the total cost and expense table. To simplify the calculation, the interest tax can be regarded as a part of the operation cost.

6. Dividend tax

For overseas investment projects invested by Chinese investors, some countries stipulate that when the owners of the projects pay dividends, they need to pay dividend tax. The calculation base of dividend tax is the annual dividend of shareholders. The dividend tax rate in most countries is 10%.

In financial evaluation, the dividend tax can be put in the table of total cost and expense. To simplify the calculation, the dividend tax can be part of the operation cost. If the annual dividend of shareholders is the product of the investment of shareholders and the profit rate payable, the dividend tax is calculated based on the dividend amount, which is relatively simple. If the dividend number of shareholders is not fixed in each year, the dividend tax shall be calculated based on the difference between the cash inflow in the cash flow statement of capital fund and the sum of project capital fund, loan principal and interest repayment, operation cost excluding dividend tax, sales tax surcharges, income tax and renovation investment.

4.5.3 Financial Evaluation Indicators

Financial evaluation indicators of the project include FIRR, FNPV, investment payback period, total investment rate of return, net profit rate of project capital, interest provision rate, debt repayment provision rate and asset liability rate.

(1) Financial internal rate of return (FIRR) is the discount rate when the accumulated present value of net cash flow in the calculation period of the project is equal to zero. The calculation formula is as follows:

$$\sum_{t=1}^{n} (CI-CO)_t (1+FIRR)^{-t} = 0$$

Where:
  FIRR—financial internal rate of return;
    CI—cash inflow, 10^4 yuan;
    CO—cash outflow, 10^4 yuan;
(CI−CO)—the net cash flow at t^{th} year, 10^4 yuan;
      t—the chronology of each year. The serial number of basis year is 1;
      n—calculation period, year.

FIRR can be calculated by using the Excel-internal function IRR (values, [guess]).

The project investment FIRR is calculated by the net cash flow in the project investment cash flow statement or the net cash flow before income tax.

The FIRR of the project capital and the FIRR of each investor shall be calculated according to the net cash flow in the corresponding financial cash flow statement.

When the FIRR is greater than or equal to the benchmark rate of return, the project scheme is financially acceptable.

According to *Construction Projects Economic Evaluation Method and Parameter* (*3rd Edition*), the pre-tax financial benchmark returns of hydropower projects, water diversion projects and water supply projects before investment and financing are 7% and 4%, respectively. And the financial benchmark returns of project capital are 10% and 6%. The capital of the enterprise shall be determined by the enterprise itself.

(2) Financial net present value (FNPV) is the sum of the present value of the net cash flow within the calculation period of the project calculated according to the set

discount rate (generally the financial benchmark rate of return). The calculation formula is as follows:

$$FNPV = \sum_{t=1}^{n} (CI-CO)_t (1+i_c)^{-t}$$

The meaning of symbols in the formula is the same as before.

FNPV can be calculated by using the Excel - internal function NPV (Rate, value1, [value 2], ...).

FNPV is an absolute indicator to evaluate the profitability of a project. It reflects the present value of the excess profit obtained by the project in addition to the profit required by the set discount rate. FNPV is equal to or greater than zero, it indicates that the profitability of the project has reached or exceeded the profitability calculated based on the set discount rate. Generally, only FNPV before income tax is calculated.

(3) The payback period (P) includes static payback period and dynamic payback period Static investment payback period is the required time generally in year, to recover the project investment based on the net income of the project. The calculation formula is as follows:

$$\sum_{t=1}^{pt} (CI-CO)_t = 0$$

The meaning of symbols in the formula is the same as before.

The static investment payback period can be calculated according to the cash flow statement. The time point when the cumulative cash flow in the cash flow statement changes from negative to zero is the static investment payback period of the project. The calculation formula is:

the static investment payback period P
= *number of years in which cumulative net cash flow begins to show positive value*
  $-1+$ *absolute value of accumulated net cash flow of last year*
  /*net cash flow value of current year*

Dynamic investment payback period is the required time generally in years, to recover the present value of project investment with the present value of net income of the project. The calculation formula is as follows:

$$\sum_{t=1}^{pt} (CI-CO)_t (1+i_c)^{-t} = 0$$

The meaning of symbols in the formula is the same as before.

dynamic investment payback period P
= *the number of years when the present value of accumulated net cash flow starts to appear positive value* $-1+$ *the absolute value of the present value of accumulated net cash flow in the previous year/the present value of net cash flow in the current year*

4.5 Financial Evaluation

The investment payback period should be calculated from the year of project construction. Otherwise, it should be noted if it is calculated from the year of project operation.

The shorter the payback period is, the higher the profitability and anti-risk ability of the project is. The criterion of investment payback period is the benchmark investment payback period, whose value can be set according to the industry level or the requirements of investors.

(4) Total investment return (ROI) is the profit level of total investment, which refers to the ratio of annual EBIT in normal years after the project reaches the design capacity or annual average EBIT to total investment (TI) during the operation period. The calculation formula is as follows:

$$ROI = \frac{EBIT}{TI} \times 100\%$$

Where:

EBIT—the annual EBIT in normal years or the annual average EBIT in the operation period, which is the total profit + annual interest expense in the profit and profit distribution statement. *the total annual profit = the total annual sales revenue − annual sales tax surcharges − annual total cost + annual interest expense*

TI—total investment of the project.

The total return on investment is higher than the reference value of the return on the same industry, indicating that the profitability meets the requirements.

(5) The net profit margin of project capital (ROE) is the profit level of project capital. It is the ratio of, annual net profit (NP) in normal years after the project reaches the design capacity or average annual net profit during the operation period, to project capital (EC). The calculation formula is as follows:

$$ROE = \frac{NP}{EC} \times 100\%$$

Where:

NP—annual net profit in normal year or annual average net profit in operation period, which is the net profit in profit and profit distribution statement;

EC—project capital.

The project capital is higher than the net profit reference value of the same industry, indicating that the profitability expressed by the net profit margin of the project capital meets the requirements.

(6) Interest provision ratio (ICR) is the ratio of EBIT to interest payable (PI) during the loan repayment period. It reflects the degree of guarantee to repay the debt interest from the perspective of the adequacy of interest payment fund source. The calculation formula is as follows:

$$\text{ICR} = \frac{\text{EBIT}}{\text{PI}}$$

Where:

 EBIT—earnings before interest and taxes. EBIT is the sum of the total profit and annual interest expense in the profit and profit distribution statement. Herein, *the total annual profit = total annual sales revenue − annual sales tax surcharges − total annual cost + annual interest expense*;

 PI—interest payable. It is the interest payable in the debt service statement.

The interest coverage rate should be calculated by year. High interest coverage rate indicates high degree of interest payment guarantee. The interest reserve rate shall be greater than 1 and shall be determined in combination with the requirements of creditors. Some creditors require that the interest reserve rate be not less than 2.

(7) DSCR is the ratio of the capital used to calculate the principal and interest repayment to the total principal and interest payable (PD) during the loan repayment period. It indicates the guarantee degree of the capital used to calculate the principal and interest repayment. The calculation formula is as follows:

$$\text{DSCR} = \frac{\text{EBITAD} - \text{Tax}}{\text{PD}}$$

Where:

 EBITDA—EBITDA plus depreciation and amortization. It can be obtained from the profit and profit distribution statement;

 Tax—income tax. It can be obtained from the profit and profit distribution statement.

According to *Construction Projects Economic Evaluation Method and Parameter (3rd Edition)*, if the project has investment to maintain operation during the operation period, the capital available for repayment of principal and interest shall be deducted from the investment to maintain operation. From the perspective of water conservancy and hydropower construction projects, due to the long depreciation period (30 – 40 years), generally there is no operation investment in the repayment period.

The debt service coverage rate should be calculated by years. The high debt service coverage rate indicates the high degree of guarantee that can be used for debt service. The debt service reserve ratio shall be greater than 1 and shall be determined in combination with the requirements of creditors. Some creditors require that the debt service reserve ratio shall not be less than 1.3.

(8) The asset liability ratio (LOAR) is the ratio of total liabilities (TL) to total assets (TA) at the end of each period. The calculation formula is as follows:

$$\text{LOAR} = \frac{\text{TL}}{\text{TA}} \times 100\%$$

Where:

TL—total liabilities at the end of the period;

TA—total assets at the end of the period.

According to *Construction Projects Economic Evaluation Method and Parameter (3rd Edition)*, a moderate asset liability ratio indicates that the enterprise is safe and stable in operation with strong financing capacity. It also shows that the risk of the enterprise and creditors is small. The analysis of the indicator should be combined with the national macroeconomic situation, the industry development trend, the competition environment, and other specific conditions. In project financial analysis, if the long-term loan is paid off, the asset liability ratio can be no longer calculated.

4.5.4 Financial Form

The financial evaluation form of water conservancy and hydropower construction project includes project investment plan and fund-raising form, total cost estimation form, loan repayment plan form, profit and profit distribution form, financial plan cash flow form, project investment cash flow form, project capital cash flow form, investors' cash flow form and balance sheet. The financial statements in *Regulation for Economic Evaluation of Water Conservancy Construction Projects* (SL 72 -2013) and the "*Specification on Economic Evaluation of Hydropower Project*" (DL/T 5441 -2010) are slightly different, which are introduced as follows. See "Chapter 9" for financial evaluation examples of hydropower projects. See "Chapter 10" for financial evaluation examples of hydropower projects.

1. Table of project investment use plan and fund-raising

The table of total investment use plan and fund-raising includes the total amount of construction investment (fixed-asset investment), interest during construction, working capital, project capital, debt capital and annual flow. It is an important table for financial analysis. The total investment in water conservancy project consists of fixed-asset investment and interest during construction. The working capital is listed separately. Whereas, the total investment in hydropower project consists of construction investment, interest during construction and working capital. Water conservancy projects, and hydropower projects are prepared according to Table 4.10. and Table 4.11.

Table 4.10   Relation between on-grid price and benchmark price in different periods of Sichuan Province (water conservancy project)   unit: 10^4 yuan

Serial number	Item	Total	Construction period					
			1	2	3	4	5	...
1	Total investment							
1.1	Fixed-asset investment							

Continued

Serial number	Item	Total	Construction period					
			1	2	3	4	5	…
1.2	Interest during construction period							
2	Working capital							
3	Fund raising							
3.1	Capital fund							
3.1.1	For fixed-asset investment							
	××party							
	…							
3.1.2	For working capital							
	××party							
	…							
3.2	Debt fund							
3.2.1	For fixed-asset investment							
	××loan							
	××bond							
	…							
3.2.2	For interest during construction period							
	××loan							
	××bond							
	…							
3.2.3	For working capital							
	××loan							
	××bond							
	…							
3.3	Other funds							
	…							

Note: For overseas water conservancy and hydropower projects financed from China, the interest during the construction period also includes overseas loan premium, loan management fee, commitment fee and handling fee.

Table 4.11　Project total investment use plan and fund raising (hydropower project)　　unit: 10^4 yuan

Serial number	Item	Total	Construction period					
			1	2	3	4	5	…
1	Total investment							
1.1	Construction investment							
1.2	Interest during construction period							

4.5　Financial Evaluation

Continued

Serial number	Item	Total	Construction period					
			1	2	3	4	5	...
1.3	Working capital							
2	Fund raising							
2.1	Project investment							
2.1.1	For construction investment							
2.1.2	For working capital							
2.1.3	For interest during construction period							
2.2	Debt fund							
2.2.1	For construction investment							
2.2.2	For working capital							
2.2.3	For interest during construction period							
2.3	Other capital							

Note: 1. If there are many kinds of loans or bonds, they should be listed separately when necessary.
2. Generally, the interest during the construction period can include other financing expenses. For example, the interest during the construction period of overseas water conservancy and hydropower projects financed from China also includes overseas loan insurance premium, loan management fee, commitment fee and handling fee.

In Table 4.11, the interest during the construction period, overseas investment and loan insurance fee, commitment fee, loan management fee and handling fee can be calculated by using the formula in Section 4.2.

The project capital is generally invested in a certain proportion of the construction investment (fixed - asset investment) or total investment by year. For overseas investment projects, it is generally calculated as a certain proportion of the EPC price.

2. Table of total cost estimation

Generaly, the total cost of water conservancy and hydropower construction projects is estimated by the production factor estimation method. It can be estimated by using the calculation method in Section 4.3 and prepared according to Table 4.12 and Table 4.13.

Table 4.12　　　　Total cost estimation (water conservancy project)　　　unit: 10^4 yuan

Serial number	Item	Total	Calculation period						
			1	2	3	4	5	6	...
1	Annual operation cost								
1.1	Material fee								
1.2	Fuel and power cost								
1.3	Repair fee								

Continued

Serial number	Item	Total	Calculation period						
			1	2	3	4	5	6	...
1.4	Staff salary								
1.5	Management fee								
1.6	Reservoir fund								
1.7	Water resources fee								
1.8	Other cost								
1.9	Fixed asset insurance premium								
2	Depreciation cost								
3	Amortization cost								
4	Financial cost								
4.1	Long-term loan interest								
4.2	Short-term loan interest								
4.3	Working capital loan interest								
4.4	Other financial cost								
5	Total cost								
5.1	Fixed cost								
5.2	Variable cost								

Table 4.13　　Total cost estimation (the factor of production method for hydropower project)　　unit: 10^4 yuan

Serial number	Item	Total	Calculation period						
			1	2	3	4	5	6	...
1	Costs of raw materials, fuel, and power								
2	Salary and welfare								
3	Repair fee								
4	Water resources fee								
5	Insurance premium								
6	Reservoir fund								
7	Other costs								
8	Operation cost (1+2+3+4+5+6+7)								
9	Depreciation cost								
10	Amortization cost								
11	Interest expense								
12	Total cost								

4.5 Financial Evaluation

Continued

Serial number	Item	Total	Calculation period						
			1	2	3	4	5	6	...
	Including: variable cost								
	Fixed cost								

Note: 1. The specific cost composition in the table is highly policy oriented and should be implemented in accordance with the relevant provisions of economic evaluation.
2. The total cost can also be estimated according to the production cost plus period cost method.
3. Some hydropower stations should be included in part of the power transmission and transformation, which should be considered in the cost analysis. It can be incorporated into the corresponding costs, or listed separately according to the needs, and reflected in other tables.

Financial cost and interest cost can be obtained from the loan repayment schedule.

For water conservancy and hydropower construction projects, due to its long production period, the income often can not meet the requirements of the total cost in the early stage of operation. To avoid negative value, the total cost table can be set with the depreciation cost and actual depreciation cost. The depreciation cost is calculated according to *Section 4.3.1*. The actual depreciation cost mainly considers the profit when the project before reaching the design capacity. During that period, the total profit is a negative value. Reduce the depreciation cost to avoid negative total profit. Thereafter, the total profit is 0. And the corresponding depreciation cost is the actual depreciation cost. For projects with low yield in the early stage of water conservancy projects, even if the actual depreciation cost is 0, the total profit is still less than 0. At this time, the actual depreciation cost is 0, which cannot be less than 0.

3. Loan repayment schedule

The repayment schedule of loan principal and interest reflects the repayment of loan principal and interest payment of each year in the calculation period. It is used to calculate the interest coverage rate, debt coverage rate and other indicators, and analyze the debt paying ability. The loan repayment schedule can be prepared according to Table 4.14 and Table 4.15. There two tables are the basic forms in financial calculation.

Table 4.14   Plan of repayment of principal and interest (water conservancy project)   unit: 10^4 yuan

Serial number	Item	Total	Construction period						
			1	2	3	4	5	6	...
1	Loan and repayment of principal and interest								
1.1	Accumulated principal and interest at the beginning of the year								
1.1.1	Principal								
1.1.2	Interest								
1.2	Loan of this year								

Continued

Serial number	Item	Total	Construction period						
			1	2	3	4	5	6	...
1.3	Accrued interest of this year								
1.4	Principal repayment of this year								
1.5	Interest payment of this year								
2	Source of repayment fund								
2.1	Undistributed profit								
2.2	Depreciation cost								
2.3	Amortization cost								
2.4	Other capital								
2.5	Interest expense included in cost								
Calculation indicators	Interest coverage ratio/%								
	Debt service coverage ratio/%								

Table 4.15　　Loan repayment schedule (hydropower project)　　unit: 10^4 yuan

Serial number	Item	Total	Construction period						
			1	2	3	4	5	6	...
1	Loan								
1.1	Loan balance at the start of the period								
1.2	Repayment of principal and interest at the current period								
	Including: principal repayment								
	interest repayment								
1.3	Loan balance at the end of the period								
2	Bond								
2.1	Bond balance at the start of the period								
2.2	Repayment of principal and interest at the current period								
	Including: principal repayment								
	interest repayment								
2.3	Closing balance of bond								
3	Total borrowings and bonds								
3.1	Opening balance								

4.5 Financial Evaluation

Continued

Serial number	Item	Total	Construction period						
			1	2	3	4	5	6	...
3.2	Repayment of principal and interest at the current period								
	Including: principal repayment								
	interest repayment								
3.3	Closing balance								
Calculation indicators	Interest coverage ratio/%								
	Debt service coverage ratio/%								

Note: If there are multiple loans or bonds, they should be listed separately if necessary.

Repayment of principal and interest can be calculated according to the formula in Section 4.3. The undistributed profit can be obtained from "Table of Profit and Profit Distribution". The depreciation cost and amortization can be obtained from "Table of Total Cost Estimation". The depreciation cost is the actual depreciation cost, and the interest cost is the medium and long-term loan interest in "Table of Total Cost Estimation". The interest reserve ratio and debt service reserve ratio shall be calculated according to the formula in Section 4.5.3.

The depreciation cost in the loan repayment fund is equal to the product of the actual depreciation and the depreciation loan repayment ratio. If there is no information, the depreciation loan repayment ratio and amortization loan repayment ratio can be calculated as 90%. The calculation method of amortization in loan repayment fund is the same as depreciation.

4. Table of profit and profit distribution

Table of Profit and Profit Distribution reflects the sales, total cost, total profit, etc. of each year during the calculation period. Also, it indicates the distribution of income tax and after-tax profit, used for the calculation of the total return on investment and the net profit margin and other indicators. Table of Profit and Profit Distribution. Also, these two tables, i.e., Table 4.16 and Table 4.17 are basic forms in financial calculation.

Table 4.16   Profit and profit distribution (water conservancy project)   unit: 10^4 yuan

Serial number	Item	Total	Construction period						
			1	2	3	4	5	6	...
	Water supply amount/$10^4 m^3$								
	Water supply price/(yuan/m^3)								
	On-grid energy/(10^4 kW·h)								
	On-grid price/[yuan/(kW·h)]								
1	Sales income								

Continued

Serial number	Item	Total	Construction period						
			1	2	3	4	5	6	...
1.1	Power generation income								
1.2	Water supply income								
1.3	Other income								
2	Sales tax and surcharges								
3	Total cost								
4	Subsidy income								
5	Total profits								
6	Make up for the loss of the previous year								
7	Taxable income								
8	Income tax								
9	Profit after tax								
10	Undistributed profit at the beginning of the period								
11	Profit available for distribution								
12	Withdrawal of statutory surplus reserve								
13	Distributable profit								
14	Profit payable by different investors								
	××party								
	...								
15	Undistributed profit								
16	BEIT (total profit+interest expense)								
17	EBIT+depreciation+amortization								

Table 4.17　　**Profit and profit distribution (water conservation project)**　　unit: 10^4 yuan

Serial number	Item	Total	Calculation period						
			1	2	3	4	5	6	...
1	Sales income								
1.1	Effective energy								
1.2	Effective capacity								
1.3	On-grid price (energy price)								
1.4	Capacity price								
2	Deductible VAT amount								
3	Sales tax and surcharges								

4.5 Financial Evaluation

Continued

Serial number	Item	Total	Calculation period						
			1	2	3	4	5	6	...
4	Total cost								
5	Subsidy income								
6	Total profits								
7	Make up for losses of previous years								
8	Taxable income								
9	Income tax								
10	Net profit								
11	Undistributed profit at the beginning of the period								
12	Profit available for distribution								
13	Withdrawal of statutory surplus reserve								
14	Profit available for distribution to investors								
15	Profits distribution of different investors								
16	Undistributed profits								
17	EBIT (total profits+interest expense)								
18	EBIT+depreciation+amortization								

Note: 1. According to the nature of the enterprise and the specific situation, we can choose to increase or decrease the items in this table.
2. The generation revenue of power system based on capacity pricing is generally composed of electricity revenue and capacity revenue. When the price is only based on electricity, the generation income is calculated based on the grid price.
3. When the VAT is deductible, the sales tax and additional calculation base shall be deducted from the sales revenue.

See "Section 4.5.1" and "Section 4.5.2" for financial income and sales tax and surcharges. The calculation formula is as follows:

$$VAT\ payable = output\ tax - input\ tax$$
$$output\ tax = output \cdot tax\ rate$$

As the input tax deductible by hydropower station is very limited, the input tax is generally not considered. The calculation formula is as follows:

$$total\ profit = sales\ revenue - sales\ tax\ surcharges - total\ cost\ expense + subsidy\ revenue$$

After making up the loss of the previous year according to the enterprise income tax law, the total profit is the taxable income, and then the income tax is levied according to law. At present, the income tax rate in China is 25%.

The net profit is the total profit deducting the income tax. The distributable profit is the net profit plus the undistributed profit at the beginning of the period. After drawing the legal surplus reserve, the profit is distributed to the investors. And the remaining profit is the undistributed profit. The calculation formula is as follows:

$$EBIT = total\ profit + interest\ paid\ in\ the\ current\ year$$
$$EBITDA = EBITDA + accrued\ depreciation + amortization$$

5. Cash flow Statement of Financial Plan

Cash flow statement of financial plan is used to reflect the cash inflow, cash outflow and net cash flow generated by the investment activities, financing activities and operating activities in each year during the calculation period of the project, to investigate the balance and balance of funds, to calculate the accumulated surplus funds, and to analyze the financial viability of the project. It is an important statement representing the financial situation. Cash flow statement of financial plan can be prepared according to Table 4.18 and Table 4.19. The data in this table can be obtained from Table of Total Investment Use Plan and Fund-Raising Statement, Table of Profit and Profit Distribution, Table of Total Cost Estimation and Table of Loan Repayment Schedule.

Table 4.18　　Cash flow statement of financial plan (water conservancy project)　　unit: 10^4 yuan

Serial number	Item	Total	Construction period						
			1	2	3	4	5	6	...
1	Net cash flow of operating activities								
1.1	Cash inflow								
1.1.1	Sales income								
1.1.2	Output tax on VAT								
1.1.3	Subsidy income								
1.1.4	Other income								
1.2	Cash outflow								
1.2.1	Annual operation cost (operating cost)								
1.2.2	Input value-added tax								
1.2.3	Sales tax and surcharges								
1.2.4	Value-added tax								
1.2.5	Income tax								
1.2.6	Other outflows								
2	Net cash flow of investment activities								
2.1	Cash inflow								
2.2	Cash outflow								
2.2.1	Fixed-asset investment								
2.2.2	Renovation investment								
2.2.3	Working capital								
2.2.4	Other outflows								

4.5 Financial Evaluation

Continued

Serial number	Item	Total	Construction period					
			1	2	3	4	5	6	...
3	Net cash flow from financing activities								
3.1	Cash inflow								
3.1.1	Project capital investment								
3.1.2	Project investment loan								
3.1.3	Short-term loan								
3.1.4	Bond								
3.1.5	Working capital loan								
3.1.6	Other inflow								
3.2	Cash outflow								
3.2.1	Long-term loan repayment of principal and interest								
3.2.2	Short-term loan repayment of principal and interest								
3.2.3	Bond redemption								
3.2.4	Repayment of working capital loan principal								
3.2.5	Long-term loan interest expense								
3.2.6	Short-term loan interest expense								
3.2.7	Working capital interest expense								
3.2.8	Accrued profits (dividend distribution)								
3.2.9	Other outflows								
4	Net cash flow								
5	Accumulated surplus capital fund								

Table 4.19　　Cash flow of financial plan (water conservancy project)　　unit: 10^4 yuan

Serial number	Item	Total	Calculation period					
			1	2	3	4	5	6	...
1	Net cash flow of operating activities								
1.1	Cash flow								
1.1.1	Sales income								
1.1.2	Substituted money on VAT								
1.1.3	Subsidy income								
1.1.4	Other income								
1.2	Cash outflow								
1.2.1	Operation cost								
1.2.2	Value-added tax								

Continued

Serial number	Item	Total	Calculation period					
			1	2	3	4	5	6	...
1.2.3	Sales tax and surcharges								
1.2.4	Value-added tax								
1.2.5	Income tax								
1.2.6	Other outflows								
2	Net cash flow from investment activities								
2.1	Cash inflow								
2.2	Cash outflow								
2.2.1	Construction investment								
2.2.2	Maintain operation investment								
2.2.3	Working capital								
2.2.4	Other outflows								
3	Net cash flow from financing activities								
3.1	Cash inflow								
3.1.1	Project capital investment								
3.1.2	Construction investment loan								
3.1.3	Working capital loan								
3.1.4	Bond								
3.1.5	Short term loan								
3.1.6	Other inflow								
3.2	Cash outflow								
3.2.1	Various interest expenses								
3.2.2	Repayment of debt principal								
3.2.3	Profit payable (dividend distribution)								
3.2.4	Other outflows								
4	Net cash flow								
5	Accumulated surplus funds								

6. Financial cash flow statement of project investment

Financial cash flow statement of project investment is viewing from the project itself, regardless of the source of investment funds. It takes the total investment as the calculation basis, reflecting the cash inflow and outflow of the whole project. It calculates FIRR, FNPV, investment payback period and other financial analysis indicators. It inspects the profitability of the project investment for the study of project decision-making. Financial cash flow statement of project investment can be prepared according to Table 4.20 and Table 4.21. The data in financial cash flow statement of project investment are from the

4.5 Financial Evaluation

basic table. The data can be obtained from "Table of Profit and Profit Distribution" and "Table of Total Cost Estimation".

Table 4.20　　Cash flow of all project investment (water conservancy project)　　unit: 10^4 yuan

Serial number	Item	Total	Calculation period						
			1	2	3	4	5	6	...
1	Cash inflow								
1.1	Sales income								
1.2	Service income								
1.3	Compensation income								
1.4	Recovery of fixed asset residual value								
1.5	Recovery of working capital								
2	Income outflow								
2.1	Fixed-asset investment								
2.2	Working capital								
2.3	Operation cost								
2.4	Sales tax and surcharges								
2.5	Renovation investment								
3	Net cash flow before income tax								
4	Accumulated net cash flow before income tax								
5	Adjustment of income tax								
6	Net cash flow after income tax								
7	Net cash flow after accumulated income tax								

Calculation indicators　　　　　　　　　　Before income tax　　　After income tax

　　FIRR of project investment
　　FNPV of project investment ($i_c =$　　%)
　　Payback period of project investment/a

Table 4.21　　Financial cash flow of project investment (hydropower project)　　unit: 10^4 yuan

Serial number	Item	Total	Calculation period						
			1	2	3	4	5	6	...
1	Cash inflow								
1.1	Generation sales income								
1.2	Compensation income and other incomes								
1.3	Project residual value								
1.4	Recovery of working capital								
2	Cash outflow								
2.1	Construction investment								
2.2	Working capital								

Continued

Serial number	Item	Total	Calculation period					
			1	2	3	4	5	6	...
2.3	Operation cost								
2.4	Sales tax and surcharges								
2.5	Maintenance operation investment								
3	Net cash flow before income tax								
4	Accumulated net cash flow before income tax								
5	Adjustment of income tax Adjustment of income tax								
6	Net cash flow after income tax								
7	Net cash flow after accumulated income tax								

Calculation indicators   Before income tax   After income tax
    FIRR of project investment
    FNPV of project investment ($i_c=$ %)
    Payback period of project investment/a

7. Project capital financial cash flow statement

Project capital financial cash flow statement is based on the investment amount of the investor from the point of view of project investors. It is used to reflect the cash inflow and outflow of the project capital, and calculate FIRR, FNPV and other financial analysis indicators. The goal is to assess the profitability of the project capital and provide reference to the project investors for decision making. Project capital financial cash flow statement can be prepared according to Table 4.22 and Table 4.23. The data in this table can be obtained from "Table of Profit and Profit Distribution", "Table of Total Cost Estimation" and "Table of Financial Plan Cash Flow".

Table 4.22   **Cash flow of capital**   unit: 10^4 yuan

Series number	Item	Total	Calculation period					
			1	2	3	4	5	6	...
1	Cash inflow								
1.1	Sales income								
1.2	Service income								
1.3	Compensation income								
1.4	Recovery of residual value of fixed assets								
1.5	Recovery of working capital								
2	Cash outflow								
2.1	Project capital								
2.2	Repayment of loan principal								

4.5 Financial Evaluation

Continued

Series number	Item	Total	Calculation period						
			1	2	3	4	5	6	...
	Including: long-term loan								
	short-term loan								
2.3	Loan interest payment								
	Including: long-term loan								
	short-term loan								
2.4	Annual operation cost								
2.5	Sales tax and surcharges								
2.6	Sales income								
2.7	Renovation investment								
3	Net cash flow								

Calculation indicator: FIRR of project capital.

Table 4.23　Financial cash flow statement of project capita (water conservancy project)　unit: 10^4 yuan

Series number	Item	Total	Calculation period						
			1	2	3	4	5	6	...
1	Income inflow								
1.1	Power generation sales revenue								
1.2	Subsidy income and other income								
1.3	Project residual value								
1.4	Recovery of working capital								
2	Income outflow								
2.1	Project capital								
2.2	Repayment of loan principal								
2.3	Loan interest payment								
2.4	Operation cost								
2.5	Sale taxes and surcharges								
2.6	Income tax								
2.7	Maintenance operational investment								
3	Net cash flow								

Calculation indicator: FIRR of project capital.

It should be noted that the principal repayment and interest cost of short-term loans in the capital cash flow statement are only short-term loans during the construction period. If the short-term loans generated due to the imbalance of financial revenue and expenditure during the operation period are borrowed, the short-term loans in the current year should be included in the cash inflow. The net cash in the current year is 0. At the same time, the principal and interest cost of short-term loans should be repaid in the next year in the cash outflow.

8. Financial cash flow statement of investors

From the perspective of project investors, financial cash flow statement of all investors analyzes the interest and tax based on investors' investment. It is used to show the cash inflow and outflow of investors. Also, it is used to calculate FIRR, FNPV and other financial analysis indicators. Besides, it can assess investors' profitability and provide reference for decision-making. Refer to Table 4.24 and Table 4.25 for details.

Table 4.24 Financial cash flow statement of investors (water conservancy project) unit: 10^4 yuan

Serial number	Item	Total	Calculation period						
			1	2	3	4	5	6	...
1	Cash inflow								
1.1	Actual profit								
1.2	Assets disposal of income distribution								
1.3	Rental income								
1.4	Capital transfer or use inflow								
1.5	Other income								
2	Cash outflow								
2.1	Actual capital contribution								
2.2	Expenditure on leased assets								
2.3	Other cash outflow								
3	Net cash flow								

Calculation indicator: FIRR of different investors.

Table 4.25 Financial cash flow statement of investors (hydropower project) unit: 10^4 yuan

Serie number	Item	Total	Calculation period						
			1	2	3	4	5	6	...
1	Cash inflow								
1.1	Dividend distribution								
1.2	Distribution of income from asset disposal								
1.3	Rental income								
1.4	Other cash inflow								
2	Cash outflow								
2.1	Equity investment								
2.2	Leased asset								
2.3	Other cash outflow								
3	Net cash flow								

Calculation indicator: FIRR of investors.

4.5 Financial Evaluation

9. Balance sheet

The balance sheet shows the increase and decrease of assets, liabilities, and owner's equity at the end of each year during the calculation period. It is used to calculate the asset liability ratio. Refer to Table 4.26 and Table 4.27 for details. The data in this balance sheet can be found from the above tables.

Table 4.26　　　　　Balance sheet (water conservancy project)　　　unit: 10^4 yuan

Serial number	Item	Total	Calculation period						
			1	2	3	4	5	6	…
1	Assets								
1.1	Total amount of working capital								
1.1.1	Monetary funds								
1.1.2	Debt receivable								
1.1.3	Prepaid accounts								
1.1.4	Goods in stock								
1.1.5	Others								
1.2	Projects under								
1.3	Net value of fixed assets								
1.4	Intangible assets and others								
2	Liabilities and owner's equality								
2.1	Total amount of current liabilities								
2.1.1	Short-term borrowing								
2.1.2	Accounts payable								
2.1.3	Unearned revenue								
2.1.4	Others								
2.2	Loan of project investment								
2.3	Loan of working capital								
2.4	Subtotal of debts								
2.5	Owner's equity								
2.5.1	Capital fund								
2.5.2	Capital reserve								
2.5.3	Accumulated surplus reserves								
2.5.4	Accumulated undistributed profits								

Calculation indicator: assets liability ratio (　%).

Table 4.27　　　　　　Balance sheet (water conservancy project)　　　　unit: 10^4 yuan

Serial number	Item	Total	Calculation period						
			1	2	3	4	5	6	...
1	Assets								
1.1	Total amount of working capital								
1.1.1	Working capital								
1.1.2	Accumulated surplus capital								
1.2	Project under construction								
1.3	Net value of fixed assets								
1.4	Intangible assets and others								
2	Liabilities and owner's equality								
2.1	Total amount of current liabilities								
2.2	Loan of construction investment								
2.3	Loan of working capital								
2.4	Subtotal of debts								
2.5	Owner's equity								
2.5.1	Capital fund								
2.5.2	Capital reserve								
2.5.3	Accumulated surplus reserves								
2.5.4	Accumulated undistributed profits								

Calculation indicator: assets liability ratio (　　%).

4.5.5　Financial Evaluation Calculation

The calculation period of financial evaluation is the period set for dynamic economic analysis. It includes construction period and operation period, generally in years. The construction period is the required time from formal investment to project completion and operation. For some large-scale hydropower projects, the power station units are put into operation gradually, and the period from the first unit to the completion of the project is called the initial operation period. That is, the construction period includes the initial operation period. Operation period is the time from project completion to the design life. Water conservancy projects can only bring benefits after the project completion. At the early period of project completion, the water demand is lower than the designed water supply. And this is known as the initial operation period.

Financial evaluation calculation of water conservancy and hydropower construction projects mainly includes financial evaluation indicator calculation, maximum loan capacity calculation, and water price calculation, etc.

1. Calculation of financial indicators

The calculation of financial indicators is to calculate each financial table and each financial indicator according to actual or predicted water price, electricity price, water supply and power generation after project implementation, as well as financing scheme. Specifically, it mainly includes construction investment (fixed-asset investment), interest during the construction period, working capital, total investment, water price, electricity price, water supply, on-grid energy, return loan period, FIRR on project investment, FNPV of project investment, investment payback period, FIRR on capital, NPV of capital, asset liability ratio, debt repayment ratio, interest coverage ratio, etc.

2. Estimation of maximum loan capacity

The estimation of the maximum loan capacity, special for water conservancy projects, is mainly to calculate the maximum loan amount and project capital amount and propose financing scheme according to possible the loan amount and capital amount. See Chapter 5 for details.

The estimation of the maximum loan capacity is to calculate the financial income based on the predicted water price, electricity price, water supply and power generation after project implementation. First, assume a loan proportion, and calculate each financial table and financial indicators. Then, determine the maximum loan amount (loan proportion) and the required capital amount through trial calculation.

3. Estimation of water price and electricity price

Based on the initial or determined financing scheme and financial calculation, the minimum water price or electricity price is calculated according to requirements of project owners or relevant sectors. The price needs to meet the requirements during loan repayment period, project investment FIRR or capital FIRR.

The estimation of water price and electricity price is based on the preliminary financing scheme. First, assume a water price or electricity price scheme, and calculate the financial income, financial table and financial indicators. Next, determine the minimum water price or electricity price that meets the requirements of loan repayment period, project investment FIRR or capital FIRR through trial calculation.

4.5.6 Financial Evaluation

1. Expenditure sequence of financial income

According to *Construction Projects Economic Evaluation Method and Parameter (3rd Edition)*, for projects with certain operation financial income, the sequence of income compensation expenses is as follows: compensation for staff salaries (salaries and welfare expenses), production and operation costs such as materials, payment of turnover tax, maintenance fees, repayment of loan interest, accrual of depreciation and amortization fees, repayment of loan principal, etc.

2. Financial viability analysis

Financial viability analysis is to analyze whether the project has enough net cash flow to maintain normal operation, to achieve financial sustainability. Specifically, it examines the cash inflow and outflow generated by investment, financing, and operating activities during project calculation period. Additionally, it calculated the net cash flow and accumulated surplus funds. Financial sustainability, is refluted in two aspects as follows: Firstly, the project should have sufficient net cash flow from operating activities. Secondly, the accumulated surplus funds in each year should not be negative. If there is a negative value, the short-term loan should be carried out, and the year length and amount of the short-term loan should be analyzed to further determine the financial viability of the project.

The financial viability of the project can be analyzed through the financial plan cash flow statement.

3. Solvency analysis

Debt paying ability analysis is to analyze and judge the debt paying ability of the financial entity according to the debt paying plan and balance sheet, through the indicators such as interest reserve ratio (ICR), debt paying reserve ratio (DSCR) and asset liability ratio (LOAR).

4. Profitability analysis

The profitability analysis is based on the project investment cash flow statement, capital cash flow statement, as well as profit and profit distribution statement. It can analyze the project profitability through FIRR, FNPV, investment payback period, return on investment (ROI) and return on equity investment (ROE).

5

Fund Sources and Financing Plan

According to *Regulation for Economic Evaluation of Water Conservancy Construction Projects* (SL 72 -2013), potential financial income of water conservancy construction projects should be analyzed according to project characteristics and market demand. Reasonable financing plans should be put forward through the analysis and comparison of capital structure, capital source and financing conditions. These can provide frame of reference for national, local governments and relevant investment departments to make decisions in early stage of hydropower projects. The fund sources and financing plan shall conform to current national policies, financial tax system and bank credit conditions. Calculation methods and main parameters shall be determined in accordance with the provisions of financial evaluation.

5.1 Fund Sources

The fund sources of water conservancy projects include project capital (equity capital) and project debt capital.

1. Project capital

Project capital is the amount of capital contribution subscribed by investors in the total investment of the construction project. It is non – debt capital for the construction project, and the project legal person does not bear any interest and debt. According to law, investors may enjoy the owner's equity in accordance with the proportion of capital contribution. Also, investors can transfer the capital contribution, but generally they are not allowed to withdraw it in any way.

Investors can make capital contributions in currency, as well as material object, industrial property, non – patented technology, land use right, resource exploitation right, etc. Except for the currency, others must be evaluated and valued by a qualified asset appraisal institution. Specifically, the proportion of capital contributed by industrial property rights and non – patented technologies shall generally not exceed the total capital of the project 20%.

The sources of capital include government capital at all levels, government subsidies, government investment, capital invested by project enterprises as legal persons, personal capital and other funds. At present, some water conservancy and hydropower projects use inundation compensation investment to participate in the project to form preferred shares. The preferred shares only participate in dividends, not in corporate control.

2. Project debt capital

Project debt capital is the capital obtained from financial institutions, securities markets and other capital markets in the form of liabilities in project investment. Debt capital has several characteristics: first, it is time – limited. Second, the principal and interest are repaid on schedule. Moreover, the capital cost is generally lower than equity capital, and investors' control over the enterprise will not be dispersed.

The sources of project debt capital include loans from domestic commercial banks, policy banks, foreign governments, international financial organizations, export credit, syndicated loans, corporate bonds and financial leasing. Commercial bank loan is an important channel of short –, medium –, and long – term loans for China's construction projects, with simple procedures and low cost. Policy bank loan is provided for projects implementing national industrial policies and other relevant policies. Generally, it has a long period and low interest rate. Specifically, the banks include China Development Bank, the Export – Import Bank of China and the Agricultural Development Bank of China. Foreign government loan is a kind of low interest preferential loan provided by one government to another with certain assistance or partial donation. It is long – term with low interest rate and limited use. Specifically, it includes Japan International Cooperation Bank loan, Japan energy loan, USAID loan, Canadian International Development Agency loan, German, French and other government loans. Loans from international financial organizations are provided to their member countries in accordance with the articles of association. Specifically, the loans include the International Monetary Fund (IMF), the World Bank and the Asian Development Bank (ADB). Export credit is loans provided by governments, which aim to promote the equipment exports and encourage the banks to provide loans to the exporters at home and abroad. The interest rate is usually lower than that of international commercial banks. But some certain additional fees are needed to be paid, such as management fee, commitment fee, and credit insurance fee etc. Syndicated loan is provided by a group of banks which are led by one or several banks. These banks adopt the same loan agreement and is in accordance with the commonly agreed loan plan. Generally, projects with large loan amount may adopt syndicated loan. Corporate bond is the debt right issued by enterprises based on their financial and credit situations. It is in accordance with the conditions and procedures stipulated in the *Securities Law of the People's Republic of China*, *Company Law of the People's Republic of China* and other laws and regulations. It is agreed that the principal and interest would be repaid within a certain period. Specifically, the corporate bond includes the Three Gorges bond and the railway bond, etc. Financial leasing is the debt right issued by the asset owner within a certain period. The asset is leased to the lessee for use, and the lessee shall pay a certain lease fee in installments.

At present, the main sources of debt capital for water conservancy and hydropower projects are loans from domestic commercial banks, policy banks, corporate bonds and financial leasing.

5.2 Financing Plan

First, establish financing body of the project. Financing body is the economic entity which carries out financing activities and bears financing responsibilities and risks. There are two financing methods, i.e., existing legal person financing and newly-created legal person financing, according to different financing bodies. ①Existing legal person financing means the existing legal person roles as the legal person of the project for financing activities. The existing legal person, without newly-created independent legal person, is responsible for capital raising and bears civil liability for the new project. The construction capital is composed of the internal financing, new capital and new debt capital of the existing legal person. Generally, the reconstruction and expansion project of water conservancy and hydropower project are financed with this method, i.e., the existing legal person financing. ②The newly created legal person financing means create new legal person and carry out financing activities for project construction. That is, to implement the new project, a new legal person with independent civil liability should be created for project financing and operation. The construction capital includes invested capital from project shareholders and debt capital from the project company. Generally, newly built water conservancy and hydropower projects are financed with this method, i.e., newly created legal person financing.

The analysis of financing plans is the basis of financial evaluation. Firstly, the pre-financing analysis of investment decision should be carried out. If reasonable results are obtained in pre-financing analysis, then the preliminary financing plan should be proposed after the analysis of the source channels and financing methods of project investment and working capital. Secondly, analyze the reliability of capital source, capital structure, financing cost and financing risk of the preliminary financing plan. At last, compare and determine the financing plan considering the results of the post-financing financial analysis.

According to the *Notice of the State Council on the Trial Implementation of the Capital System for Fixed Asset Investment Projects* (GF〔1996〕 No. 35): the capital ratios are different for different industrial projects. Specifically, transportation and coal projects, it is 35% or more. For iron and steel, post and telecommunications, fertilizer projects, it is 25% or more. For electric power, electromechanical, building materials, chemical industry, petroleum processing, non-ferrous, light industry, textile, commerce and other industries projects, it is 20% or more. Specifically, the capital ratios shall be determined by project approval unit when examining and approving the feasibility study report. The economic benefits of the investment project and the bank's willingness to loan and the evaluation opinions should be considered. The total investment, as the calculation base, is the sum of the fixed asset investment and the initial working capital. Specific verifications should reference to the approved dynamic budget estimate.

According to *Regulation for Economic Evaluation of Water Conservancy Construction Projects* (SL 72 -2013): Different loan proportions are designed for water conservancy projects with different main services. Specifically, the loan proportion of that mainly serves for power generation is no more than 80%. Whereas, the loan proportion of that mainly serves for urban water supply (water transfer) is no more than 65% in principle. The loan proportion of others is determined according to the results of loan capacity measurement and project situation, no more than 80%.

Project financing is usually called financing, and capital financing program is called financing plan. For water conservancy and hydropower projects, during project proposal, the financing plan should be initialized. During the feasibility study, after receiving the approval from the national or local Development and Reform Commission, the financing plan should be defined. When analyzing the financing plan, the maximum loan capacity should be firstly calculated.

The measurement of the maximum loan capacity aims at predicting financial income and expenses according to market demand and measuring the maximum loan amount and capital amount to maintain project operation under the current financial tax system and bank credit conditions.

According to *Regulation for Economic Evaluation of Water Conservancy Construction Projects* (SL 72 -2013), when calculating the loan capacity, it is necessary to calculate the cost of water and electricity products according to the cost sharing results of comprehensive utilization of construction projects. Put forward the market competitiveness of the project products after the investigation of the supply and demand of water and electricity products and the market prospect. Analyze the impact of other water and power sources on the project. Put forward the affordability of different users to water price and electricity price. The price schemes of water and electricity products shall be analyzed and proposed according to the similar market price, project cost price, profit price, user affordable price, price approved by the competent department, price agreed by both suppliers and consumers, etc. Specifically, as the following:

5.2 Financing Plan

(1) reference to the water supply (electricity) price of recent similar water conservancy projects. Predict according to the development level and planning of regional national economy, as well as the development and utilization of water resources and the supply and demand of electricity.

(2) the project cost shall be calculated according to the apportioned expenses and shall be formulated according to the cost and investment profit.

(3) the consumer's willingness and ability to pay should be considered.

(4) the price of water conservancy products should be agreed upon by both the suppliers and the consumers.

(5) the policy price approved by the competent price department or relevant government departments.

For water conservancy projects with annual sales revenue larger than the annual total cost, the loan capacity must be measured. Otherwise, if the annual sales revenue is smaller than the annual total cost, the loan capacity may not be measured. When the project annual sales revenue is larger than annual operation cost but smaller than annual total cost, the loan capacity may be analyzed according to actual situations, after considering the costs for project renewal and reconstruction and the financial condition during loan repayment.

For thewater conservancy projects to apply for national investment as capital, the financing plan that all capital is not distributed profit during the loan repayment period should be calculated. And this plan should be taken as the basic financing plan. Hereafter, other financing plans are calculated according to the capital income requirements of enterprise legal person.

In the initial period of the water conservancy projects, the water consumption is relatively small, and the water supply benefits can not be brought into full play. Besides, the early repayment pressure is large with equal principal and interest or equal principal repayment. Generally, when calculating the loan capacity, the short - term loan in the operation period does not exceed five years to control the maximum loan capacity. The net cash flow can be paid to the enterprise after the short - term financing plan is completed, when the net cash flow in the cash flow statement is larger than zero. If the net cash flow of the project is zero during the whole loan repayment period, and the profit can not be distributed to investors, then a part of the capital can be reserved to distribute profits to the financing owners every year to reduce the loan amount.

For water conservancy projects with various deb tcapital, according to debt capital use conditions, it is necessary to evaluate the impact of plans, with different debt capital structure, on project financing capacity. Results of loan capacity is provided in Table 5.1.

The project financing schemes are proposed based on feasible plans of water price, electricity price and loan repayment period.

5　Fund Sources and Financing Plan

Table 5.1　　　　　　　　　　**Results of loan capacity**

NO.	Loan period /a	Energy price /(yuan /kW·h)	Water price /(yuan /m³)	Loan capital /10⁴ yuan	Capital /10⁴ yuan	Construction period interest /10⁴ yuan	Total investment /10⁴ yuan	Project FIRR /%	Capital FIRR /%	Proportion of loan capital to fixed-assets investment/%	Note

[**Example 5.1**] This project mainly serves for water supply and power generation, and the fixedasset investment is 1733.58 million yuan. According to the investment allocation and cost calculation, the total cost water price is 0.40 yuan/m³. When the investment profit rates are given as 1%, 2%, 3% and 4%, the corresponding water prices are 0.48 yuan/m³, 0.55 yuan/m³, 0.63 yuan/m³ and 0.70 yuan/m³, respectively. The water price is 1.03 yuan/m³ meeting the FIRR of capital fund of 6%. For the current water supply project in the receiving area, raw water price is 0.10 yuan/m³. For local water supply project near the receiving area, raw water price is 0.20 – 1.0 yuan/m³. The price of domestic water in the receiving area is 1.40 – 1.70 yuan/m³. The affordable water price for urban residents is 4.5 – 9.1 yuan/m³, when the water expenditure accounts for 1% – 2% of the household income. The affordable water price at the reservoir outlet is 0.6 – 2.4 yuan/m³, considering the water supply and distribution cost, as well as the water supply cost proportion of the water transmission project.

For this project, the total cost price is 0.14 yuan/(kW·h). And the corresponding on-grid price is 0.30 – 0.23 yuan/(kW·h) at the profit rates of 6% – 8% as required. For the power stations of large and medium-sized reservoirs, which are approved by the Provincial Price Bureau in recent years, the on grid price is 0.32 – 0.36 yuan/(kW·h).

Assume the loan periods as 20 years and 25 years, and the annual loan interest rate as the currently used, long-term loan interest rate over 5 years, i.e., 5.65%. Considering the total cost water price and different profit rate water price, six plans are proposed: 0.40 yuan/m³, 0.48 yuan/m³, 0.55 yuan/m³, 0.63 yuan/m³, 0.70 yuan/m³ and 1.03 yuan/m³. The proposed on-grid price is 0.32 yuan/(kW·h) and 0.36 yuan/m³. Combined plan of loan capacity of water supply price and on-grid price is provided in Table 5.2.

5.2 Financing Plan

Table 5.2 Combined plan of loan capacity of water supply price and on-grid price

Scheme	Loan period	On-grid price/[yuan/(kW·h)]	Water supply price/(yuan/m³)	Annual income /10⁴ yuan	Loan capital /10⁴ yuan	Capital /10⁴ yuan	Construction interest /10⁴ yuan	Total investment /10⁴ yuan	FIRR/% Project investment	FIRR/% Project capital	Ratio of loan capacity to fixed-asset investment /%
1	20	0.32	0.40	11670	46971	126387	6651	180009	1.38	0.49	27.1
2			0.48	12916	52578	120780	7445	180803	2.05	2.18	30.3
3			0.55	14006	57400	115958	8128	181486	2.58	1.77	33.1
4			0.63	15252	63100	110258	8936	182294	3.16	2.42	36.4
5			0.70	16342	67833	105525	9605	182963	3.63	2.97	39.1
6			1.03	21480	90793	82565	12857	186215	5.62	5.52	52.4
7		0.36	0.40	12214	51419	121939	7282	180640	1.70	0.78	29.7
8			0.48	13460	57067	116291	8081	181439	2.34	1.47	32.9
9			0.55	14550	61922	111436	8769	182127	2.86	2.05	35.7
10			0.63	15796	67494	105864	9557	182915	3.42	2.70	38.9
11			0.70	16886	72429	100929	10256	183614	3.88	3.26	41.8
12			1.03	22024	95376	77982	13506	186864	5.78	5.85	55.0
13	25	0.32	0.40	11670	54071	119287	7657	181015	1.35	0.01	31.2
14			0.48	12916	60579	112779	8578	181936	2.04	0.70	34.9
15			0.55	14006	66138	107220	9366	182724	2.58	1.32	38.2
16			0.63	15252	72600	100758	10280	183638	3.15	2.01	41.9
17			0.70	16342	78263	95095	11082	184440	3.63	2.61	45.1
18			1.03	21480	104820	68538	14843	188201	5.56	5.55	60.5
19		0.36	0.40	12214	59239	114119	8389	181747	1.70	0.26	34.2
20			0.48	13460	65631	107727	9294	182652	2.34	0.98	37.9
21			0.55	14550	71405	101953	10111	183469	2.86	1.60	41.2
22			0.63	15796	77746	95612	11010	184368	3.42	2.30	44.8
23			0.70	16886	83454	89904	11818	185176	3.88	2.91	48.1
24			1.03	22024	109864	63494	15557	188915	5.78	5.96	63.4

Considering the current water price in the receiving area and the user's psychological capacity, the recommended water price is 0.63 yuan/m³, and corresponding profit rate is 3%. The recommended on grid price is 0.32 yuan/(kW·h). The loan repayment period is 20 years. Loan principal is 631 million yuan, accounting for 36.4% of the fixed-asset investment. The interests during the construction period is 89.36 million yuan. The project capital is 1102.58 million yuan. The total investment is 1822.94 million yuan, as that of Plan 4 shown in Table 5.2. The average net cash during the loan repayment period (the 5th to the 20th years) is 20.54 million yuan, and the net cash of the 5th to the 9th year is zero.

According to the recommended water price and electricity price, under the condition of meeting the loan repayment requirements, the loan principal is 631 million yuan. Project capital is 1102.58 million yuan. The FIRR of the project capital is 2.42%, less than 8% of the minimum return requirement of the enterprise. Therefore, it is difficult to attract enterprise investment. Government subsidy is needed, and it can be injected as capital fund, but this part of capital fund does not need to pay payable profits, that is, the FIRR of government capital fund is zero. According to the FIRR of 6%, 7%, 8%, 9% and 10% of the capital of the enterprise, the capital of the enterprise that can be absorbed is 522.23 million yuan, 428.9 million yuan, 353.93 million yuan, 294.39 million yuan and 245.88 million yuan, respectively. Details are shown in Table 5.3.

Table 5.3　　　　　　　　Capital components of reservoir project

Scheme	On-grid price/[yuan/(kW·h)]	Water supply price/(yuan/m³)	Loan capital /10⁴ yuan	Project capital /10⁴ yuan	Company capital FIRR/%	Company capital /10⁴ yuan	Government capital /10⁴ yuan	Proportion of enterprise capital to project capital/%
1	0.32	0.63	63100	110258	6	52223	58035	47.4
2	0.32	0.63	63100	110258	7	42890	67517	38.9
3	0.32	0.63	63100	110258	8	35393	75014	32.1
4	0.32	0.63	63100	110258	9	29439	80968	26.7
5	0.32	0.63	63100	110258	10	24588	85819	22.2

According to requirements of the investment enterprise, the FIRR of the enterprise capital is 8%. The corresponding capital that can be absorbed is 353.93 million yuan. The needed government capital is 750.14 million yuan. Therefore, the enterprise capital takes up 32.1% of the project capital. Assume that the average net cash during the loan repayment period is 20.54 million yuan, then the average profit margin payable can reach 5.8%.

Project financing plan: the loan principal is 631 million yuan. The government capital is 750.14 million yuan. The enterprise capital is 3539.31 million yuan. The loan interest during the construction period is 89.36 million yuan. The fixed-asset investment is 1733.58 million yuan. The total investment is 1822.94 million yuan.

6

Uncertainty Analysis

According to *Regulation for Economic Evaluation of Water Conservancy Construction Projects* (SL 72 -2013), uncertainty analysis and risk analysis shall be carried out for water conservancy construction projects. Specifically, these are required to evaluate the economic reliability and possible risk of the project, put forward the risk early warning and prevention countermeasures, to serve the investment decision. Uncertainty analysis includes sensitivity analysis and break-even analysis. Specifically, the sensitivity analysis is mainly used in the uncertainty analysis of national economic evaluation and financial evaluation of water conservancy construction projects. Whereas, the break-even analysis is used for financial evaluation of important water conservancy projects, which are having financial benefits. At last, for particularly important water conservancy construction projects, risk analysis should be carried out.

According to *Specification on Economic Evaluation of Hydropower Project* (DL/T 5441 -2010), most of the data used in the economic evaluation of hydropower construction projects have some uncertainty because they are from prediction and estimation. To analyze the impacts of uncertain factors on the economic evaluation index and estimate the potential risks, the uncertainty and risk analysis of the future power market and the project construction, and possible impacts should be taken as the focus. For hydropower construction projects, the uncertain analysis is mainly the sensitivity analysis. Sometimes, further risk analysis should be further carried out accordingly.

6.1 Sensitivity Analysis

6.1.1 The Purpose and Significance of Sensitivity Analysis

Sensitivity analysis is a kind of uncertainty analysis methods. Firstly, it finds out the sensitive factors of the economic benefit index of the investment project. Then, the degree of impacts and sensitivity of the identified factors are analyzed and calculated. Finally, the risk tolerance of the project is evaluated.

The purposes of sensitivity analysis are as follows:

(1) Find out the sensitive factors that affect the economic benefits of the project. Analyze why the sensitive factors varies and provide the basis for further uncertainty analysis (such as probability analysis).

(2) Study the range or limit value of the uncertain factors, such as the economic benefit value. Analyze and judge the project ability to bear the risk.

(3) The sensitivity of several schemes is compared, so that the insensitive investment schemes can be selected in case of similar economic benefits.

According to the number of uncertainty factors at each change, sensitivity analysis can be divided into single-factor and multi-factor sensitivity analysis. In the former (single-factor sensitivity analysis), only one factor is changed at a time to estimate the impact of this single factor on project benefits. In the latter (multi-factor sensitivity analysis), two or more factors are changed at the same time to estimate the impacts of multi-factors on project benefits. In the national economic evaluation and financial evaluation, the single-factor sensitivity analysis is usually adopted to find out the key sensitive factors. However, single-factor and multi-factor sensitivity analysis can be carried out simultaneously, if necessary.

6.1.2 Methods and Steps of Sensitivity Analysis

1. Select uncertain factors

Firstly, the uncertain factors should be identified, and the degree of deviation from the normal mean should be determined. Uncertain factors are the basic factors having impacts on the project benefits. Currently, only the important uncertain factors are considered in sensitivity analysis of national economic evaluation and financial evaluation. Those include: construction investment (fixed-asset investment), power generation and water supply benefits, loan interest rate, loan repayment period, yield rate, construction period, exchange rate, etc.

2. Determine the degree of uncertainty

In sensitivity analysis, the adverse and favorable impacts of uncertain factors are usually considered simultaneously to investigate the comprehensive impacts on benefit indicators. The sensitivity analysis table or chart is produced.

Generally, the change of most uncertain factors is expressed by percentage, such as $\pm 5\%$, $\pm 10\%$, $\pm 15\%$, etc. For others that are not suitable to be expressed by percentage, such as loan repayment period and construction period, the method of extending or shortening the period can be adopted.

3. Selected analysis indicators

According to the characteristics of the project, sensitivity analysis shall analyze the impact of one or multiple indicators on the main economic evaluation indices. These indices include construction investment (fixed-asset investment), project benefit, price of main inputs and outputs, construction period and exchange rate. The most basic analysis indicators include internal rate of return, net present value (NPV), product price, etc. Other evaluation indicators such as investment payoff period can also be selected according to the project situation. If necessary, sensitivity analysis can be carried out for two or more indicators simultaneously.

Generally, in the financial analysis of water conservancy and hydropower projects, the selected analysis indicators in sensitivity analysis are power station electricity price, water-supply price, financial internal rate of return of project investment, etc. Whereas, in the economic analysis, the selected analysis indicators are economic net present value (ENPV) or economic internal rate of return (EIRR).

4. Identify sensitivity analysis indicators

(1) Sensitivity coefficient. Sensitivity coefficient (S_{AF}) is the ratio of the percentage change of project benefit index to the percentage change of uncertainty factor. High sensitivity coefficient indicates that the project benefit is highly sensitive to the uncertain factors. Therefore, attention should be paid to these uncertain factors. The sensitivity coefficient is expressed as follows:

$$S_{AF} = \frac{\Delta A/A}{\Delta F/F}$$

Where:

S_{AF}—sensitivity coefficient of evaluation index A to uncertain factor F;

$\Delta F/F$—change rate of uncertain factor F;

$\Delta A/A$—when the uncertain factor F changes ΔF, the corresponding change rate of evaluation index A;

$S_{AF} > 0$, indicates that the evaluation index changes in the same direction as the uncertain factors; $S_{AF} < 0$, indicates that the evaluation index changes in the opposite direction with the uncertain factors. The higher the $|S_A|$, the higher the sensitivity coefficient.

(2) Critical point. The critical point is the limit value of the uncertain factor. It is the critical value at which the project becomes infeasible from feasible, or the change rate at which IRR = the basic rate of return or NPV = 0. When the uncertain factor is cost item, it is the increased percentage. When the uncertain factor is the benefit item, it is the reduced percentage. When the change of uncertain factors exceeds the critical point, the project benefit index will be lower than the benchmark value indicating that the project turns infeasible from feasible.

The critical point is correlated with the set benchmark yield. With the increase of the set benchmark yield, the critical point will become lower.

[**Example 6.1**] For X Investment Plan, the initial investment is 30 million yuan. The annual net income is 4.8 million yuan. The service life is 10 years. The benchmark yield is 10%. The ending residual value is 2 million yuan. The single factor sensitivity analysis of NPV when the main parameters of initial investment, annual net income, life cycle and benchmark rate of return change separately is carried out. The sensitivity analysis is conducted as follows:

(1) Determine the economic evaluation index—NPV.

(2) Assume the variation rate of each factor is k. The variation range is ± 30%. The interval is 10%.

(3) Calculate the NPV_j ($j=1, 2, 3, 4$) when each factor, including initial investment, annual net income, life cycle and benchmark yield, changes separately. The calculation formula is as follows:

$$NPV_1 = 480(P/A, 10\%, 10) + 200(P/F, 10\%, 10) - 3000(1+k)$$
$$NPV_2 = 480(1+k)(P/A, 10\%, 10) + 200(P/F, 10\%, 10) - 3000$$
$$NPV_3 = 480[P/A, 10\%, 10(1+k)] + 200[P/F, 10\%, 10(1+k)] - 3000$$
$$NPV_4 = 480[P/A, 10\%(1+k), 10] + 200[P/F, 10\%(1+k), 10] - 3000$$

See Table 6.1 for calculation results.

Table 6.1 Project net present value (NPV) when the parameters change separately

Parameter \ Variation	-30%	-20%	-10%	0	10%	20%	30%
Variation of initial investment	927	627	327	27	-273	-573	-873
Annual net	-858	-563	-268	27	321	616	911
Life length variation	-560	-346	-151	27	170	334	468
Basic rate of return	479	313	170	27	-98	-224	-436

According to the data in Table 6.1, the sensitivity analysis diagram is drawn. x-coordinate represents the change rate of parameter, and the vertical coordinate represent NPV, as shown in Figure 6.1.

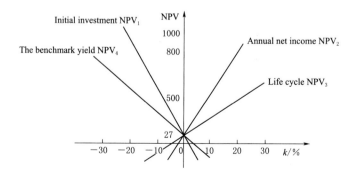

Figure 6.1 Sensitivity analysis

(4) Identify the sensitive factors and judge the risk of the scheme. On the sensitivity analysis diagram, find out the intersection point of each sensitivity curve and the x-coordinate. The parameter value at this point is the critical value that makes NPV=0.

The intersection point between the sensitive curve and the x-coordinate is about 0.98%, and the corresponding initial investment is:
$$K = 3000 \times (1 + 0.98\%) = 30.29 \text{ million yuan}$$

That is, when the initial investment increases to 30.29 million yuan, the NPV will drop to zero. It indicates that the initial investment must be controlled below 30.29 million yuan before the scheme is feasible.

The intersection point of annual net income and x-coordinate is about -1%. The annual net income (M) which makes the scheme feasibility is:
$$M \geqslant 480 \times (1 - 1\%) = 4.75 \text{ million yuan}$$

The intersection point between the life cycle and the x-coordinate is about -5%. The life cycle (N) that makes the project feasible is:
$$n \geqslant 10 \times (1 - 5\%) = 9.5 \text{ year}$$

The intersection point between the benchmark yield and the x-coordinate is about 4.8%. The benchmark yield which makes the scheme feasible is:
$$i \leqslant 10\% \times (1 + 4.8\%) = 10.48\%$$

Comparing the critical change rate k_j of each factor and the shape of the sensitive curves, we can learn that if the absolute value of the critical change rate is small, the sensitive curve is steeper, and the change of the corresponding parameters has a greater impact on the NPV.

[Example 6.2] The sensitivity analysis of the national economic evaluation of × Hydropower Station project is carried out. The uncertain factors include construction investment and valid energy. The change range is between ±10%. The analyzed index is EIRR. The benchmark yield at the critical point is 8%. The sensitivity coefficient and the critical point of each uncertain factor are calculated as follows:

(1) Construction investment increased by 10%.
$$\Delta A/A = (0.08 - 0.095)/0.095 = -0.158$$
$$S_{AF} = -0.158/0.1 = -1.58$$

$S_{AF} < 0$, indicates that the evaluation index and uncertain factor change in the opposite direction. That is, the construction investment increases, but the EIRR decreases.

(2) Available power increased 10%.
$$\Delta A/A = (0.114 - 0.095)/0.095 = 0.2$$
$$S_{AF} = 0.2/0.1 = 2.0$$

$S_{AF} > 0$, indicates that the evaluation index and the uncertain factor change in the same direction. That is, the valid energy increases, and the economic internal rate of return also increases.

6 Uncertainty Analysis

The calculation results are shown in Table 6.2.

Table 6.2   Sensitivity analysis of ×× hydropower station

Serial number	Uncertain factors	Floating range/%	EIRR/%	Sensitivity coefficient	Critical point/%
		Basic plan	9.5		
1	Construction investment	+10	8.0	−1.58	10
		−10	10.5	1.05	
2	Effective power	+10	11.4	2.0	−9.6
		−10	7.8	−1.79	

Comparing the absolute value of the sensitivity coefficient of the above two uncertain factors, we can see that the sensitivity coefficient of the effective power is greater than that of the construction investment. It indicates that the effective power has a greater impact on the project benefit index than the construction investment.

6.2 Break-even Analysis

6.2.1 The Concept of Break-even Analysis

Break-even analysis is also known as volume-cost-profit analysis. It is a method to calculate the break-even points and analyze the balanced relation among the project output, cost and profit according to the product output (sales volume), fixed cost, variable cost, product price and tax, in the normal year.

In the production and sales of an engineering project, there is at least one break point, i.e., BEP, between the profit and loss. The break-even points are determined by three factors: fixed cost, variable cost and unit product price. Break-even analysis includes linear and non-linear break-even analysis. The linear break-even analysis is generally carried out for the water conservancy and hydropower construction projects.

Break-even points can be expressed in various forms, such as product output, product price, specific variable cost and fixed cost. The mostly used are product output and capacity utilization rate.

6.2.2 The Function and Basic Assumption of Break-even Analysis

Through the break-even analysis, the break-even points can be found. In addition, the adaptability and anti-risk ability of the project, to the production change caused by the market, can be investigated. In terms of product output and capacity utilization rate, the smaller the break-even point the larger the enterprise's adaptability to the market demand, the stronger the anti-risk ability. The lower the break-even point in terms of the product price, the greater the enterprise's adaptability to the decline of market price and the stronger the anti-risk ability. Break even analysis is generally only used in financial evaluation.

The following assumptions are made in break-even analysis:
(1) The output of the product is equal to the sales volume.
(2) The variable cost per unit product remains unchanged.
(3) The sales unit price of unit product remains unchanged.
(4) The products can be converted into a single type.
(5) All used data are from normal production year.

6.2.3 Calculation Method of Break-even Analysis

The break-even point can be calculated using following equations:

BEP(production capacity utilization rate)
= annual total fixed cost/(annual sales revenue
 − annual total variable cost − annual sales tax and surcharges)
×100%

BEP(output)
= annual total fixed cost(unit product price − variable cost
 − unit product sales tax and surcharges)×100%
= BEP(production capacity utilization rate)
 ×design production capacity

BEP(product price)
= annual total fixed cost/(production capacity utilization rate
 + unit variable cost + unit product sales tax and surcharges)

Besides, the break-even point can be obtained by graphic method. For example, as shown in Figure 6.2, the intersection point of the sales revenue line and the total cost line is the break-even point (BEP).

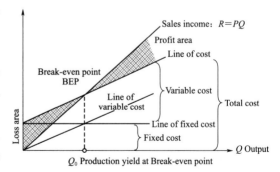

Figure 6.2 Break-even analysis (product output)

[**Example 6.3**] The designed annual production of X Project is 120000 t. The sales price is 770 yuan/t. The product tax is 150 yuan/t. The variable cost per unit product is 250 yuan. The total annual fixed cost is 15 million yuan. What is the production yield and production capacity utilization rate at the break-even point?

Solution:

Production yield at break-even point:
$$Q(\text{BEP}) = 15000000/(770-150-250) = 40540.54 \ (t)$$

Production capacity utilization rate:
$$R(\text{BEP}) = 40540.54/120000 \times 100\% = 33.78\%$$

[**Example 6.4**] × Hydropower Station takes 3 years for the construction. In the 4th year, it starts the operation. The revenue includes power generation revenue (402.9 million kW·h since the 4th year) and CDM revenue (13.71 million yuan during the 4th–10th year of the calculation period). The total cost in each year of the calculation period is different. Specifically, it is 140.89 million yuan in the 4th year, including the operation cost of 17.84 million-yuan, depreciation cost of 45.4 million-yuan, loan interest of 77.66 million yuan. To reach breakeven, the electricity price of the 4th year shall be 0.379 yuan/(kW·h). During the loan repayment period, the loan interest is reduced year by year, and the total cost is also reduced year by year. The break-even price is 0.379–0.213 yuan/(kW·h). After the loan repayment (during the 26th–33rd year), the break-even price is 0.193–0.195 yuan/(kW·h). The break-even price of each year is shown in table 6.3.

Table 6.3 Break-even price of each year

Serial number	Item	operation period												
		4	5	6	7	8	9	10	11	12	13	14	15	16
I.	Total cost	14089	13932	13764	13583	13390	13183	12962	12725	12472	12200	11909	11598	11265
1	Operation cost	1784	1784	1784	1784	1784	1784	1784	1784	1784	1784	1784	1784	1784
2	Depreciation cost	4540	4540	4540	4540	4540	4540	4540	4540	4540	4540	4540	4540	4540
3	Interest expenditure	7766	7608	7440	7260	7066	6860	6638	6402	6148	5877	5586	5275	4942
3.1	Long-term loan interest expenditure	7761	7604	7435	7255	7062	6855	6634	6397	6143	5872	5581	5270	4937
3.2	Working capital loan interest expenditure	5	5	5	5	5	5	5	5	5	5	5	5	5
II.	income	14089	13932	13764	13583	13390	13183	12962	12725	12472	12200	11909	11598	11265
1	CDM income	1371	1371	1371	1371	1371	1371	1371	0	0	0	0	0	0
2	Generation revenue	12913	12753	12582	12399	12204	11994	11770	12901	12644	12368	12074	11758	11421
3	Business tax and surcharges	194	192	190	187	185	182	179	175	172	168	164	160	155
III.	Break-even price													
1	Energy price excluding tax	0.324	0.320	0.315	0.311	0.306	0.301	0.295	0.323	0.317	0.310	0.303	0.295	0.286
2	Energy price including tax	0.379	0.374	0.369	0.364	0.358	0.352	0.345	0.378	0.371	0.363	0.354	0.345	0.335

Continued

Serial number	Item	operation period																
		17	18	19	20	21	22	23	24	25	26	27	28	29	30	31	32	33
Ⅰ.	Total cost	10909	10527	10238	9806	9345	8850	8320	7753	7146	6496	6504	6513	6522	6532	6542	6553	6564
1	Operation cost	1784	1784	1903	1909	1915	1922	1928	1936	1943	1951	1960	1968	1978	1987	1998	2008	2020
2	Depreciation cost	4540	4540	4540	4540	4540	4540	4540	4540	4540	4540	4540	4540	4540	4540	4540	4540	4540
3	Interest expenditure	4585	4204	3795	3358	2890	2388	1852	1278	663	5	5	5	5	5	5	5	5
3.1	Long term loan interest expenditure	4581	4199	3791	3353	2885	2384	1847	1273	658	0	0	0	0	0	0	0	0
3.2	Working capital loan interest expenditure	5	5	5	5	5	5	5	5	5	5	5	5	5	5	5	5	5
Ⅱ.	income	10909	10527	10238	9806	9345	8850	8320	7753	7146	6496	6504	6513	6522	6532	6542	6553	6564
1	CDM income	0	0	0	0	0	0	0	0	0	0	0	0	0	0	0	0	0
2	Generation revenue	11059	10672	10379	9942	9473	8972	8435	7860	7245	6585	6594	6603	6612	6622	6632	6643	6655
3	Business tax and surcharges	150	145	141	135	129	122	115	107	99	90	90	90	90	90	90	90	91
Ⅲ.	Break-even price																	
1	Energy price excluding tax	0.277	0.268	0.260	0.249	0.238	0.225	0.211	0.197	0.182	0.165	0.165	0.166	0.166	0.166	0.166	0.167	0.167
2	Energy price including tax	0.324	0.313	0.304	0.292	0.278	0.263	0.247	0.231	0.213	0.193	0.193	0.194	0.194	0.194	0.195	0.195	0.195

6.3 Risk Analysis

After the operation of water conservancy and hydropower projects the possibility of actual financial and economic befits from the expected values because of uncertain factors is termed as risk. Through identifying risk factors, using qualitative or quantitative analysis methods, risk analysis estimates the possibility of various risk factors and their impacts on the project, reveals the key risk factors affecting the project, and puts forward corresponding countermeasures. Through the information feedback of risk analysis, improve or optimize the design scheme to reduce the project risk.

Based on the related study at home and abroad, the existing risks of water conservancy and hydropower projects can be summarized as follows:

1. Benefit risk

The main benefits of water conservancy and hydropower projects are from water supply and power generation. The benefit risks include project actual production risk and market risk.

(1) Project actual production risk. Due to the problems such as the length of the in-

flow runoff series adopted or the representativeness of hydrological station data, there are some deviations between the actual inflow runoff and the design results after the project operation. If the actual inflow is less, the benefit of the project will be affected to a certain extent and there is a certain benefit risk.

In the design process, we should strengthen the collection and analysis of meteorological and hydrological data, strive to obtain accurate and reliable hydrological data, so that the project benefits can be implemented.

(2) Market risk. The market risk of hydropower project mainly refers to that the power supply scope of the power station fails to meet the original design requirements, the load forecast error of the power grid, the construction of other power stations ahead of time, which causes the power grid unable to absorb the on-grid power of the power station according to the design requirements and leads to the electricity unsalable. The market risk of water conservancy project mainly refers to the lag of supporting projects, the slow growth of water users, and the water failure to meet the original design requirements Unsalable.

2. Risks induced by insufficient preliminary survey

Generally, water conservancy and hydropower projects are built in mountain and canyon areas. Geological survey, topographic survey, underwater topographic survey of river channel, site selection of construction materials, etc. are required to provide basic data for the design of hydraulic structures and the construction of underground projects (water tunnel, foundation treatment, underground powerhouse, etc.). If the depth of preliminary work, the number of boreholes is insufficient or the survey position deviates, the reliability of the design results will be affected. In the early stage of the work, it is necessary to carry out in-depth exploration work on key parts of main buildings to reduce the restrictive impact of the early stage of the survey work on the project construction and keep the risk under control.

3. Risk of construction period

Water conservancy and hydropower projects are constructed in the field and site specific. Exposed to various uncertainties and risk factors, such as climate, traffic, material and equipment supply, the projects usually suffer some risks at the construction period. Particularly, for projects with a large proportion of underground works, the risk of construction period should be paid special attention to.

The construction of underground tunnel and underground powerhouse is greatly affected by topographic conditions, geological conditions, groundwater, surface flood and so on, which affects the construction progress of the projects with a large proportion of underground projects. In the process of engineering construction, it is necessary to arrange the construction period and construction team reasonably according to the relevant engineering experience in order to meet the requirements of the construction period.

4. Risk of project construction cost control

Project construction cost overrun is often caused by many factors. The main causes are as follows:

(1) Design change, mainly including the project design plan adjustment and project quantity change due to the change of geological conditions.

(2) Price rising and price adjustment, including the adjustment of unit price of raw materials such as cement, steel, oil and labor, and the transportation expenses adjustment of highway, railway and ocean freight, etc.

5. Risk of exchange rate

With more and more domestic companies participating in the construction of foreign water conservancy and hydropower projects, exchange rate risk has become one of the unavoidable risks.

Foreign projects are generally constructed in the form of EPC (engineering procurement construction) or BOT (Build Operate Transfer) contracts. The main equipment and building materials are purchased in China. The construction managers and main technical workers are all employees of Chinese enterprises. The project investment estimate is denominated in Chinese Yuan and converted into US dollars at the current exchange rate. The income is generally denominated in US dollars or local currency. The exchange rate is related to local social situation, and changes in policies and regulations.

7

Economic Comparison of Project Schemes

Scheme comparison is an important part of project economic evaluation. For the design standards, project scale, project layout and main design schemes of water conservancy projects, various schemes should be proposed. First, carry out the economic calculation. Next, carry out detailed demonstration and comparison in combination with other factors, and select the optimal scheme. The economic comparison of technical schemes shall be determined according to the results of national economic evaluation. It can also be determined according to the results of financial evaluation, if there is no conflict with that of national economic evaluation.

7.1 Scheme Comparability

In comparing technical schemes, to make the schemes economically comparable and comprehensively reflect the relative economic value, these four comparable principles should be followed:

(1) Need comparability. In terms of product quantity, quality, time, location, and reliability, all the comparison schemes need to meet the needs of national economic development to the same extent.

(2) Comparability of consumption costs. The comparability of consumption costs includes:

1) From the perspective of total consumption of the system and society, the scheme cost itself, other costs related to the scheme and various costs of adjacent departments (such as raw materials, fuel, power, transportation, etc.) shall be considered in the cost of consumption of the technical schemes.

2) Only after the total consumption cost of the comprehensive utilization scheme is apportioned can it be compared with a scheme that can only meet the unilateral needs.

3) The consumption cost indexes of each technical scheme should be comparable, whether it is investment index or cost index. For example, due to the different quality and

calorific value of coal, the investment per ton of coal can not be compared. They can be compared only after converted into the investment per ton of standard coal.

(3) Price comparability. Price comparability includes:

1) The product price of the technical schemes shall be comparable.

2) The prices of various products adopted in the calculation of the consumption expenses of the technical schemes, especially the prices of energy, raw materials and transportation in the cost expenses, shall be comparable.

3) Different technical schemes should adopt the price level of corresponding period.

(4) Time comparability. Time comparability includes:

1) The economic comparison of different technical schemes should be based on the same calculation period. If the calculation periods are different, two following schemes can be adopted for adjustment: ① Extend the service life of the shorter normal operation period scheme and consider the cost of equipment renewal and transformation. ② Shorten the service life of the longer normal operation period scheme and recover the corresponding residual value of fixed assets at the end of the calculation period.

2) Mind that different technical schemes have different economic impacts overall national economy due to the differences input of human, material, financial, transportation, natural resources and time.

3) Time comparability is for both major projects and supporting projects.

7.2 Relationship among Schemes

Generally, multiple alternative schemes, but not single scheme is provided for project investors. The global optimum of multiple schemes is pursued, but not the local optimum of a single scheme. There are three types of relationship among the alternative schemes: mutual exclusion type, independent type and layer mixing type.

(1) Mutually exclusive scheme. The characteristic of mutual exclusion is that there is mutual exclusion among schemes. Only one of these schemes can be selected. Once this scheme is selected, other schemes must be abandoned.

(2) Independent scheme. The characteristic of independent schemes is that they are compatible with each other. If conditions support, the advantageous schemes can be chosen arbitrarily. These schemes can coexist, and the investment, operation cost and income are additive.

(3) Layer mixing scheme. The characteristic of layer mixing type is that the scheme has two levels. The high level scheme is composed of a group of independent projects, and each independent project is realized by several mutually exclusive schemes.

The comparison and selection of water conservancy and hydropower construction projects are basically mutually exclusive. Specifically, this is introduced in next sections.

7.3 Comparison Methods of Mutually Exclusive Schemes

According to specific conditions and project capital, the economic comparison of mutually exclusive schemes can adopt the methods of benefit comparison, cost comparison, minimum price, minimum cost, maximum effect and increment analysis.

7.3.1 Benefit Comparison Method

When the cost and benefit of the comparison and selection scheme can be monetized, the benefit comparison and selection method can be adopted under the condition of no capital constraints. The method of benefit comparison includes the method of internal rate of return of difference investment, the method of net present value and the method of net annual value. Generally, it can be used for comparison of normal water level, dead water level, reservoir dam site, installed capacity, installed capacity of power station, unit type of power station, rated head of power station, tunnel diameter of Headrace Tunnel (or size of Headrace channel), etc.

When the investment and benefit of each alternative are different, and the calculation period is the same, the method of internal rate of return of difference investment and the method of net present value should be mainly used. If the calculation period of the comparison scheme is different, the net annual value method can be used.

7.3.1.1 Method of IRR of Differential Investment

The IRR of differential investment is the discount rate when the present value of the difference between the net benefit flows of each year in the calculation period of the two schemes for scheme comparison is equal to zero.

The IRR of differential investment is calculated according to the following formula:

$$\sum_{t=1}^{n} [(B-C)_2 - (B-C)_1]_t (1+\Delta \text{IRR})^{-t} = 0$$

Where:

ΔIRR—differential investment, ΔEIRR or ΔFIRR;

$(B-C)_2$—the annual net benefit flow of the scheme with large investment present value;

$(B-C)_1$—the annual net benefit flow of the scheme with small investment present value.

The differential investment economic internal rate of return (ΔEIRR) or financial internal rate of return (ΔFIRR) for scheme comparison is expressed by the discount rate when the present value of the difference between the net benefit flows of each year in the calculation period of the two schemes is equal to zero. When the IRR of the differential investment is greater than or equal to the social discount rate (or the set benchmark rate of return), the scheme with a large present value of the investment is preferred. Otherwise, the scheme with a small present value of the investment is preferred. Schemes should be

7.3 Comparison Methods of Mutually Exclusive Schemes

compared in order of the present value of the investment.

[Example 7.1] The construction period, the initial operation period and the normal operation period are the same for two alternative schemes. However, both the costs and benefits are different. Capital investment is provided in Table 7.1. Calculate the IRR of the differential investment between the two schemes and determine the better economic scheme.

Table 7.1   Scheme comparison results

Chronological order	Scheme I			Scheme II			Difference in annual net benefits of Schemes I, II
	Benefit	Cost	Net benefit	Benefit	Cost	Net benefit	
t	B	C	$(B-C)_1$	B	C	$(B-C)_2$	$(B-C)_2 - (B-C)_1$
(1)	(2)	(3)	(4)	(5)	(6)	(7)	(8)
1	0	900	−900	0	1050	−1050	−150
2	100	500	−400	80	600	−520	−120
3	200	20	180	250	35	215	35
4	200	20	180	250	35	215	35
⋮	⋮	⋮	⋮	⋮	⋮	⋮	⋮
31	200	20	180	250	35	215	35
32	200	20	180	250	35	215	35

Apply the Excel-internal function IRR () to the 8th column. The calculated IRR of the differential investment is 11.7%. It is larger than the social discount rate of 8%. Therefore, Scheme II is better.

7.3.1.2 Net Present Value Method

The NPV method refers to comparing the economic or financial NPV of each alternative scheme, and the scheme with large NPV is preferred. When comparing NPV, the same discount rate should be used, which is applicable to the comparison between schemes with the same calculation period. NPV is calculated as follows.

$$\text{NPV} = \sum_{t=1}^{n} (B - I - C' + S_v + W)_t (1 + i_s)^{-t}$$

Where:
  B—the benefit;
  I—the sum of fixed asset investment and current capital;
  C'—the annual operation cost;
  S_v—the residual value of fixed assets recovered at the end of calculation period, 10^4 yuan;
  W—the working capital recovered at the end of calculation period, 10^4 yuan;
  n—the calculation period in terms of year;

i_s—social (set) discount rate;

NPV—net present value.

[Example 7.2] According to the data in "Example 7.1", the social discount rate is 8%, and the method of net present value is used to select the economically better scheme.

Solution: The calculation base point is at the beginning of the 1st year. Apply the Excel-internal function NPV (Rate, value1, [value2] …) to the 4th column and the 7th column, respectively. The calculated NPV of Scheme Ⅰ and Scheme Ⅱ is 5.61 million yuan and 6.57 million yuan, respectively. The NPV of Scheme Ⅰ is smaller than that of Scheme Ⅱ. Therefore, Scheme Ⅱ is better. This result is consistent with the result of IRR of differential investment.

7.3.1.3 Net Annual Value Method

In the net annual value method, the NAV values of alternative schemes are compared. The scheme with larger NAV is the optimal scheme. The same discount rate should be used when comparing the NAV. And the NAV is calculated as follows:

$$\text{NAV} = \left[\sum_{t=1}^{n} (B - I - C' + S_v + W)_t (1 + i_s)^{-t} \right] (A/P, i_s, n)$$

Where:

NAV—the net annual value, 10^4 yuan;

$(A/P, i_s, n)$—the capital recovery factor.

Others are the same as before.

It can be seen from the formula that the economic NAV is the product of the ENPV and the capital recovery factor.

[Example 7.3] Use the data in "Example 7.1". The social discount rate is 8%. Select the economical scheme by use of the net annual value method.

Solution: The calculation base point is at the beginning of the 1st year. Apply the Excel-internal function NPV (Rate, value1, [value2], …) to the 4th column and the 7th column, respectively. The calculated NPV of Scheme Ⅰ and Scheme Ⅱ is 5.61 million yuan and 6.57 million yuan, respectively. Then, apply the Excel-internal function PMT (Rate, Nper, Pv, [Fv], [Type]). The calculated NAV of Scheme Ⅰ and Scheme Ⅱ is 0.49 million yuan and 0.57 million yuan. The NAV of Scheme Ⅰ is smaller than that of Scheme Ⅱ. Therefore, Scheme Ⅱ is better. This result is consistent with Examples 7.1, and 7.2.

7.3.2 Cost Comparison Method

When the benefits of the alternative schemes are the same, or more disgusted, they can meet the same requirements and achieve the same purpose, but it is difficult to estimate their benefits, the cost present value method or the annual cost method can be used for comparison. The scheme with small cost present value or annual cost value is the one

7.3 Comparison Methods of Mutually Exclusive Schemes

with good economic effect. The cost present value method is only applicable to the same calculation period of each scheme, and the annual cost method can be applicable to the same calculation period or different calculation periods. For example, the cost comparison method can be adopted for the comparison of water supply engineering lines, gravity flow water supply scheme and water lifting scheme of water supply engineering.

7.3.2.1 Cost Present Value Method

The cost present value method is to compare the cost present value (PC) of each alternative, and the scheme with small cost present value is preferred. The present value of the cost is calculated as follows:

$$PC = \sum_{t=1}^{n}(I + C' - S_v - W)_t (P/F, i_s, t)$$

Where:

PC—the present value of cost, 10^4 yuan.

Others are the same as before.

[**Example 7.4**] There are two schemes for the urban water supply construction project in × area. Scheme I is to build a reservoir to supply water to the water supply target by gravity. The construction investment is 500 million yuan, and the annual operation cost is 10 million yuan. Scheme II is to build a water pumping station. The construction investment is 250 million yuan, and the annual operation cost is 35 million yuan. The water supply benefits of the two schemes are the same. For economic comparison of the two schemes, the cost present value method is adopted. The social discount rate is 8%. Results of scheme comparison in Table 7.2.

Table 7.2   Results of scheme comparison

Chronological order	Scheme I			Scheme II		
	Investment	Annual operation cost	Total cost	Investment	Annual operation cost	Total cost
(1)	(2)	(3)	(4)	(5)	(6)	(7)
1	50000		50000	25000		25000
2		1000	1000		3500	3500
3		1000	1000		3500	3500
4		1000	1000		3500	3500
⋮	⋮	⋮	⋮	⋮	⋮	⋮
30		1000	1000		3500	3500
31		1000	1000		3500	3500

Solution: Apply the Excel - internal function NPV (Rate, value1, [value2], ...) to the 4th column and the 7th column, respectively. And the calculated present cost value of Scheme I and Scheme II is 567.2 million yuan and 596.32 million yuan, respectively. The present cost value of Scheme I is smaller than that of Scheme II. For alternative schemes, the smaller present cost value, the better the economic effect. Therefore, Scheme I is better.

7.3.2.2 Annual cost method

The annual cost method is to compare the equal annual cost (AC) of each scheme, and the scheme with the lowest AC is the best. AC is calculated as follows:

$$AC = \left[\sum_{t=1}^{n}(I+C'-S_v-W)_t(P/F,i_s,t)\right](A/P,i_s,n)$$

Where:

AC—the annual cost value, 10^4 yuan.

Other symbols are the same as before.

It can be seen from the above formula that the annual cost value (AC) is the product of the present cost value and the fund recovery factor, so the annual cost value method and the present cost value method are the same.

[**Example 7.5**] Use the data in "Example 7.4". The social discount rate is 8%. Select the economical scheme using the annual cost method.

Solution: at the beginning of the first year, apply the Excel – internal function NPV (Rate, value1, [value2], …) to the 4th column and the 7th column, respectively. The calculated NPV of Scheme I and Scheme II is 567.2 million yuan and 596.32 million yuan, respectively. Then, apply the Excel – internal function PMT (Rate, Nper, Pv, [Fv], [Type]). The calculated NAV of Scheme I and Scheme II are 50.38 million yuan and 52.97 million yuan, respectively. The NAV of Scheme I is smaller than that of Scheme II. Therefore, Scheme I is better. The result is consistent with that of Example 7.4.

7.3.3 Other Comparison Methods

Other comparison methods can also be adopted in scheme comparison and selection. However, they are not commonly used in economic evaluation of water conservancy and hydropower construction projects. Herein, only the method definitions are introduced, and no examples are given.

1. Minimum price method

If the product of the project is a single product and the product output of each scheme is different, the lowest price method shall be adopted. For each scheme, the lowest price of the product is calculated when NPV=0. The scheme having the lowest product price is preferred.

2. Maximum effect method

When the effect of each scheme is the same, the benefits can not be monetized, and the cost – effectiveness analysis is required, the minimum cost method shall be adopted. The scheme with the least cost is the best. Under the same cost, the maximum effect

7.3 Comparison Methods of Mutually Exclusive Schemes

method shall be used, and the scheme with the largest effect is the best. When the effect of each scheme is significantly different from the cost, the incremental analysis method shall be adopted to compare the cost difference and effect difference between the two alternative schemes.

3. Incremental analysis method

Incremental analysis method is also known as difference analysis method. In this method, the same part of the two schemes are omitted, and only the different parts are compared. It evaluates the economic effect of incremental investment by calculating the difference cash flow. It is generally used to compare the relative economy of the two schemes.

The incremental analysis method should be adopted when both the effect and cost of the two schemes are significantly different. It should be measured according to whether the difference ratio is reasonable.

The incremental effect cost ratio is calculated as follows:

$$R_{\Delta E/\Delta C} = \frac{E_2 - E_1}{C_2 - C_1}$$

Where:

$R_{\Delta E/\Delta C}$ — the incremental effect cost ratio;

E_2, E_1 — the scheme with large effect and small effect, respectively;

C_2, C_1 — the scheme with large cost and small cost in the calculation period, respectively, and expressed by present value or annual value.

The benchmark index $R_{E/C}$ is the minimum requirement of the acceptable effect cost ratio of the project. It is determined comprehensively according to the national economic situation, industrial characteristics and the ratio level of the previous similar projects.

When the incremental effect cost ratio is greater than or equal to the benchmark index ($R_{\Delta E/\Delta C} \geq R_{E/C}$), the scheme with higher cost is the one with better economic effect. Otherwise, the scheme with lower cost is the one with better economic effect. Schemes should be compared in order of cost from small to large.

8

Cost Allocation

Generally, water conservancy projects in China have multiple purposes such as flood control, power generation, irrigation, water supply, shipping, and ecology. Some even benefit multiple areas. Therefore, the investment and annual operation cost of the multi-purpose water conservancy projects shall be allocated and calculated among the beneficiary sectors or areas according to reasonable principles and appropriate methods.

The aims of cost allocation of multi-purpose water conservancy projects are: ①provide basis for the calculation of economic benefit indicators in different sectors, and the determination of reasonable development target and scheme. ②coordinate the requirements of different beneficiary sectors or areas, select, and determine economic and reasonable project scale and technical parameters. ③provide reference for the preparation of national construction plan, the arrangement of investment plan, and fund raising of construction funds as well as annual operation cost. ④provide basis for accounting and reasonably determining the cost and price of water conservancy products such as water supply and power generation.

8.1 Investment Classification of Multi-purpose Water Conservancy Construction Projects

The multi-purpose water conservancy project generally includes reservoir, barrage, flood discharge facilities, water diversion facilities, power station, water diversion system, navigation facilities and fish passing facilities. In investment allocation, the investment is usually divided into common investment and special investment. Besides, there are dual-purpose projects and compensation projects.

Special investment is the multi-purpose engineering facilities specially built for a certain sector in water conservancy projects, such as power stations and water diversion systems, navigation facilities, water diversion facilities, etc. Shared investment is the engineering facilities used by all beneficiary sectors, such as reservoirs, barrages, flood discharge facilities, etc. Dual-purpose project is the project or facility that only serves a certain beneficiary sector or region, but also has the common function. For example, riverbed power plant is not only the special building of the power station, but also has the common function of water retaining. Compensation project is some engineering facilities built to maintain or compensate the interests of some sectors or regions. After the reservoir construction, the diversion capacity of the downstream facilities for water diversion without dam will be reduced, so compensation projects need to be built.

8.2 Principle of Cost Allocation

Principles of cost allocation for multi-purpose water conservancy construction projects are listed in *Regulation for Economic Evaluation of Water Conservancy Construction Projects* (SL 72 -2013).

(1) The cost in the project of common purpose for each function shall be reasonably determined through cost allocation.

(2) For engineering facilities serving only a few functions, these functions can be regarded part first in the allocation of the total cost. Then allocate the shared cost among these functions.

(3) For the engineering facilities that mainly serve a specific function and are an indispensable part of the project, as well as have certain effect on other functions, the alternative cost of works for common purpose shall be allocated among the benefit functions. The part exceeding the cost of works for common purpose shall be paid by the specific function.

(4) The special cost for a function in the multi-purpose water conservancy construction project shall be paid by the function itself.

(5) If a function is damaged after the construction of the project, the cost of taking remedial measures to restore the original functions shall be paid by all beneficial functions. The increased project cost beyond the original effective energy shall be borne by the function.

8.3 Investment Allocation in Special Projects and Projects of Common Purpose

The project investment in multi-purpose water conservancy project is estimated in a unified way. In investment allocation, special project investment needs to be extracted. The following are the functions of the typical buildings of multi-purpose water conservancy project.

1. Dams

The reservoir dam has the function of water retaining. For the reservoir with flood control function, the added reservoir capacity and dam height only for flood control function is considered as the investment in flood control function. It shall be determined by hydraulic engineering quantity.

2. Powerhouse of hydropower station

For hydropower station at the dam toe and diversion type hydropower station, the investment in civil engineering and mechanical and electrical engineering belongs to the power generation sector. This part investment is the special investment in power generation. For riverbed hydropower station, the civil engineering part is not only the special project, but also plays the role of water retaining. For interpretation wall, it is allocated between the power generation and water retaining according to a certain proportion, and the remaining is the special investment in power generation.

3. Water diversion project

Headworks buildings and control equipment shall be the special cost of water supply sector. If the investment in main and branch diversion canals (pipes) involves multiple beneficiary sectors, it shall be divided into several sections. It shall be allocated according to the common investment of these beneficiary sectors by water supply proportion or other methods.

4. Navigation building

Construct water conservancy and hydropower projects on non-navigable rivers or river reaches. The cost in the navigation buildings, regardless of their scale, shall be listed as the special investment in shipping sector.

If the scale of navigable buildings, constructed as part of the newly constructed water conservancy and hydropower projects, does not exceed the original navigable grade or navigable capacity of the river course, the navigable buildings are considered as the compensatory projects. Accordingly, this shall be treated as investment in project of common purpose. Otherwise, the cost of the excess part shall be treated as special investment in the shipping sector. And the part of the original capacity corresponding to the investment shall be the compensation project and the project of common purpose.

5. Other facilities

Fishway facilities belong to compensation projects. Accordingly, the investment is invested in works for common purpose, and shall be allocated among all beneficiary sectors.

In the calculation of the investment in common and special projects, it is necessary to calculate the direct investment of each building. Calculate corresponding temporary project investment and other investment according to the proportion of direct investment. Then calculate the building investment.

8.4 Cost Allocation Method of Works for Common Purpose

There are many methods for the cost of works for common purpose. According to *Regulation for Economic Evaluation of Water Conservancy Construction Projects* (SL 72-2013), the proposed methods include: allocation according to proportions of storage capacity, allocation according to water amount, allocation according to present value of equivalent alternative cost, allocation according to present value of benefit, separable costs-remaining benefits method (SCRB), primary and secondary allocation method, etc.

1. Allocation according to proportions of reservoir storage capacity

In this method, the cost of works for common purpose is allocated according to the size of storage capacity required by each function, and the size of storage capacity required by each function can be determined according to the runoff regulation calculation results. For projects with different monthly utilization capacity, the annual average allocation proportion can be calculated based on the monthly utilization capacity proportion. This method is suitable for reservoirs with regulating capacity requirements for each function. However, it has certain limitations for the power generation function of that after water supply and irrigation regulation.

[Example 8.1] Consider a project mainly serves for flood control and water supply. The special investment of water supply project is 40 million yuan, and the investment of shared project is 2200 million yuan. According to the regulation calculation, the total storage capacity of the reservoir is 132.96 million m^3, including 80 million m^3 for benefit storage capacity, 7 million m^3 for dead storage, 15 million m^3 for flood control, and no combined storage. The investment allocation shall be made according to the proportion of flood control and beneficial storage capacity. The calculation is as follows:

The allocation proportion of flood control function is:

($flood\ control\ storage\ capacity$)/($dead\ storage\ capacity + benefit\ storage\ capacity + flood\ control\ storage\ capacity$) = 1500/(700+8000+1500) = 14.7%

The allocation proportion of water supply function is:

($dead\ storage\ capacity + benifit$)/($dead\ storage\ capacity + benifit\ storage\ capacity + flood\ control\ storage\ capacity$) = (700+8000)/(700+8000+1500) = 85.3%

The investment of flood control function sharing project is $22 \times 14.7\% = 323.4$ million yuan

The investment of water supply function sharing project is $22 \times 85.3\% = 1876.6$ million yuan.

Considering the special investment of water supply function of 40 million yuan, the total investment of water supply function sharing project is 1916.6 million yuan. See

Table 8.1 for the results of investment allocation.

Table 8.1 Result of investment allocation based on storage capacity method

Item	Utilization capacity/$10^4 m^3$	Utilization capacity proportion/%	Allocated investment of sharing project/10^8 yuan	Investment in special project/10^8 yuan	Total allocated investment/10^8 yuan	Allocated investment proportion in fixed assets/%
Flood	1500	14.7	3.234		3.234	14.4
Water supply	8700	85.3	18.766	0.4	19.166	85.6
Total	10200	100.0	22.0		22.4	100.0

2. Allocation according to water consumption

In this method, the cost of works for common purpose is allocated according to the proportions of the designed water supply of different water supply objects. Additionally, different assurance rates of water supply should be considered. This method is suitable for water conservancy projects mainly serves for water supply or the proportion sharing among several water users. The following are examples of investment allocation between different water users and different regions.

[**Example 8.2**] The project serves for irrigation, urban water supply, etc. The irrigation water supply is 40 million m^3, and the water supply assurance rate is 75%. The urban water supply is 50 million m^3, and the water supply assurance rate is 95%. The fixed asset investment of the project is 1590 million yuan. Specifically, the investment in irrigation project is 40 million yuan, the investment in urban water supply project is 50 million yuan, and the investment in shared project is 1500 million yuan. Accordingly, the calculation results are as follows:

The irrigation allocation proportion is:

(*irrigation water supply* × *irrigation assurance rate*)/(*irrigation water supply* × *irrigation assurance rate* + *urban water surpply* × *urban water supply assurance rate*) = (4000×75%)/(4000×75%+5000×95%) = 38.7%

The allocation proportion of urban water supply is:

urban water supply × *urban water supply assurance rate*/*irrigation water supply* × *irrigation assurance rate* + *urban water supply* × *urban water supply assurance rate* = (5000×95%)/(4000×75%+5000×95%) = 61.3%

Irrigation allocation investment is 15×38.7% = 580.5 million yuan

The allocation investment of urban water supply is 15×61.3% = 915.5 million yuan

See Table 8.2 for investment allocation results.

8.4 Cost Allocation Method of Works for Common Purpose

Table 8.2 Investment allocation result by using water consumption proportion method

Item	Water supply amount/ $10^4 m^3$	Assurance rate/%	Allocation proportion /%	Allocated investment in sharing project/10^8 yuan	Investment in special project/ 10^8 yuan	Total allocation investment/ 10^8 yuan	Proportion of investment in fixed assets/%
Irrigation	4000	75	38.7	5.805	0.4	6.205	39.0
Urban water supply	5000	95	61.3	9.195	0.5	9.695	61.0
Total			100	15.0	0.9	15.9	100.0

[**Example 8.3**] For urban water supply in areas B, C, D and E, a reservoir, i.e., Reservoir A is needed to be built. Besides, a canal passes from Reservoir A to E via B, C and D. The total water supply of Reservoir A is 400 million m³. The allocated water of areas B, C, D and E is 60 million m³, 90 million m³, 50 million m³ and 200 million m³, respectively. Given that the reservoir investment is 2540 million yuan, and the investment of lines AB, BC, CD and DE are 730 million yuan, 960 million yuan, 860 million yuan and 1500 million yuan, respectively. Calculate the investment allocation. Location of the canal and water receiving areas of the water supply project in Figure 8.1. The results are in Table 8.3.

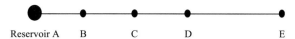

Figure 8.1 Location of the canal and water receiving areas of the water supply project

Table 8.3 Investment allocation result by using water consumption proportion method

Area		B	C	D	E	Total
Water supply amount/10^4 m³		0.6	0.9	0.5	2.0	4.0
Allocated investment in Section A	Allocated proportion/%	15.00	22.50	12.50	50.00	100.00
	Allocated investment/10^8 yuan	3.80	5.72	3.18	12.70	25.40
Allocated investment in Section AB	Allocated proportion/%	15.00	22.50	12.50	50.00	100.00
	Allocated investment/10^8 yuan	1.10	1.64	0.91	3.65	7.30
Allocated investment in Section BC	Allocated proportion/%		26.47	14.71	58.82	100
	Allocated investment/10^8 yuan		2.54	1.41	5.65	9.60
Allocated investment in Section CD	Allocation proportion/%			20.00	80.00	100.00
	Allocated investment/10^8 yuan			1.72	6.88	8.60
Allocated investment in Section DE	Allocated proportion/%				100	100
	Allocated investment/10^8 yuan				15.00	15.00
Total apportioned investment		4.90	9.90	7.22	43.88	65.90
Proportion/%		7.44	15.02	10.96	66.58	100.00

3. Allocation according to the proportion of the present value of the cost of the optimal equivalent alternative of each function

In this method, the cost of works for common purpose is allocated based on the proportion of the present value of the cost of the optimal equivalent alternative scheme of each service function. This method is suitable for the multi-purpose project where the benefit is not easy to calculate. The disadvantage is that the alternative scheme of each function needs to be determined. For the function with multiple alternative schemes, the optimal alternative scheme is taken. Some of the optimal equivalent alternatives only replace the shared project. For example, if the dike replaces the reservoir for flood control, only the shared project is replaced. Comparatively, some others replace both the shared project and the special project. For example, if the alternative thermal power method is used to replace the hydropower station, the shared investment and the special investment of the hydropower station are both replaced. For the first case, the allocation is relatively simple. See Example 8.4 for details. For the second case, there are two ways to deal with it. One is to allocate the fixed-asset investment of water conservancy project (the sum of the investment of shared project and special project) according to the proportion of the cost present value of the optimal equivalent alternative scheme, as shown in Table 8.5 of Example 8.5. The other is to allocate the proportion of the difference between the cost present value of the optimal equivalent alternative scheme and the cost present value of the special project, as shown in Table 8.6 of Example 8.5. For simplification, the first method can be used. However, for the first method, the rationality analysis is required.

[**Example 8.4**] Consider a comprehensive utilization project serves for flood control, power generation and water supply. The fixed-asset investment of the project is 1530 million yuan. Specifically, the special investment in flood control, power generation and water supply projects are 130 million yuan, 340 million yuan, and 50 million yuan, respectively. Additionally, the investment in shared project is 1010 million yuan. The present value of the cost in the optimal equivalent alternative shared project of flood control, power generation and water supply functions are 430 million yuan, 750 million yuan and 540 million yuan, respectively. The cost is allocated according to the proportions, and the calculation results are shown in Table 8.4.

Table 8.4　　Result of investment allocation by using equivalent substitution method

Item	Present value of alternative project cost/ 10^8 yuan	Allocated investment in sharing project/%	Allocated investment in sharing project/ 10^8 yuan	Investment in special project/ 10^8 yuan	Total allocated investment/ 10^8 yuan	Allocated investment proportion of fixed assets/%
Flood control	4.3	25.0	2.53	1.3	3.83	25.0
Power generation	7.5	43.6	4.40	3.4	7.8	51.0

8.4 Cost Allocation Method of Works for Common Purpose

Continued

Item	Present value of alternative project cost/10^8 yuan	Allocated investment in sharing project/%	Allocated investment in sharing project/10^8 yuan	Investment in special project/10^8 yuan	Total allocated investment/10^8 yuan	Allocated investment proportion of fixed assets/%
Water supply	5.4	31.4	3.17	0.5	3.67	24.0
Total	17.2	100.0	10.1	5.2	15.3	100.0

[**Example 8.5**] The same as "Example 8.4", given that the present value of the costs of flood control, power generation and water supply functions, optimal equivalent to the key project (including shared project and special project) is 430 million yuan, 750 million yuan and 540 million yuan, respectively. Allocate the sum of the investment in project of common purpose and the special project according to the proportion of the present value of the cost of the optimal equivalent alternative scheme. See Table 8.5 for the results of investment allocation. On the other hand, allocate the investment according to the proportion of the difference between the cost present value of the optimal equivalent alternative scheme and the cost present value of the special project. See Table 8.6 for the results of investment allocation.

Table 8.5 Results of investment allocation by using equivalent substitution method

Item	Present value of alternative project cost/10^8 yuan	Allocation proportion/%	Allocated investment/10^8 yuan	Investment in special project/10^8 yuan	Allocated investment in sharing project/10^8 yuan
Flood control	4.3	25.0	3.8	1.3	2.5
Power generation	7.5	43.6	6.7	3.4	3.3
Water supply	5.4	31.4	4.8	0.5	4.3
Total	17.2	100.0	15.3	5.2	10.1

Table 8.6 Results of investment allocation by using equivalent substitution method

Item	Present value of alternative project cost/10^8 yuan	Present value of special project cost/10^8 yuan	Difference in present value of alternative project cost and special project cost/10^8 yuan	Allocated Investment proportion of shared project/%	Allocated investment in shared project/10^8 yuan	Investment in special project/10^8 yuan	Total allocated investment/10^8 yuan	Allocated investment proportion of fixed assets/%
Flood control	4.3	0.92	3.38	25.00	2.52	1.3	3.82	25.00
Power generation	7.5	2.41	5.09	37.65	3.80	3.4	7.20	47.07
Water supply	5.4	0.35	5.05	37.35	3.77	0.5	4.27	27.93
Total	17.2	3.68	13.52	100.00	10.1	5.2	15.3	100.00

4. Allocation according to the proportion of present value of benefits available for each function

In this method, the cost of works for common purpose is allocated based on the present value or annual value of the benefits of the multi-purpose water conservancy project during the calculation period. The present value of the benefits can be determined as the benefits of all functional projects in the national economic evaluation. Since the benefits of special projects are included, the fixed-asset investment of water conservancy projects (the sum of the investment of shared projects and special projects) can be allocated according to the present value ratio of each functional benefit. After the investment allocation, the rationality analysis shall be carried out to make sure that the allocated investment is not less than the special investment.

[**Example 8.6**] Consider a project which mainly serves for power generation and irrigation. The power generation benefit is calculated by shadow price method. The present value is 907.91 million yuan. At the same time, the irrigation benefit is calculated by sharing coefficient method. The present value is 273.93 million yuan. The investment in project construction is 1025.51 million yuan. Specifically, the special investment in power generation, irrigation headworks and shared projects is 422.19 million yuan, 59.67 million yuan, and 543.65 million yuan, respectively. See Table 8.7 for the result of investment allocation. The investment in each function allocation is greater than that of the special projects. It indicates that the allocation result is appropriate.

Table 8.7　Result of investment allocation using benefit-present value method

Item	Present value of benefits/10^4 yuan	Apportionment ratio/%	Share investment/10^4 yuan	Investment in special project/10^4 yuan	Allocated investment in sharing project/10^4 yuan
Irrigation	27393	23.2	23792	5967	17825
Power generation	90791	76.8	78759	42219	36540
Total	118184	100.0	102551	48186	54365

5. Cost allocation by using separable costs-remaining benefits method (SCRB)

The method is to divide the total cost of the multi-purpose water conservancy project into two parts, i.e., the separable costs and the remaining cost of each function. Then, use the proportion of the remaining benefit to share the remaining cost. The separable cost plus the remaining cost shared is the total cost of this function. Although this method is complex, it is relatively scientific and widely used in the United States, Japan, India and some countries in Europe. It is also used in cost allocation of China's Three Gorges Project and other multi-purpose water conservancy projects.

The following are the terms of the separable costs-remaining benefits method (SCRB):

(1) Separable costs. For multi-purpose water conservancy projects, the total in-

8.4 Cost Allocation Method of Works for Common Purpose

vestment in project construction including a certain function is different from that of project construction excluding the function. The investment in the latter is generally smaller than that in the former. The difference between the two is called the separable costs of the function. For example, a multi-purpose key project is composed of functions including flood control, power generation, water supply, etc. The separable cost of the flood control function is the difference between the total investment in key construction composed of three functions including flood control function and the total investment in key construction composed of two functions excluding flood control. Similarly, the separable cost of power generation and water supply can be calculated.

(2) Remaining cost. The sum of separable investment of each function is generally smaller than the total investment in key project construction, and the difference between them is called remaining cost.

(3) Remaining benefits. The so-called remaining benefits refer to the difference between the appropriate cost and the separable cost. The appropriate cost is equal to the small value between the benefit of the function and the cost of the optimal equivalent alternative. In other words, the appropriate cost is the upper limit of the shared cost, and the separable investment is the lower limit. The difference between the two is the remaining benefit.

The separable costs-remaining benefits method (SCRB) needs more basic data. Specifically, the data include: ①the total investment, annual operation cost, and annual benefit of multi-purpose water conservancy project. ②separable investment, separable annual operation cost and benefit of each service function. ③optimal equivalent alternative project investment and annual operation cost of each service function.

[Example 8.7] Consider a water conservancy project that mainly serves for flood control, water supply, and power generation, etc. The investment in shared project is 2500 million yuan. See Table 8.8 for the results of investment allocation based on the separable costs-remaining benefits method (SCRB).

Table 8.8 Results of investment allocation by using the separable costs-remaining benefits method (SCRB)
unit: 10^4 yuan

Serial number	Item	Flood control	Water supply	Power generation	Total	Note
(1)	Fixed-asset investment				250000	Provided by preliminary calculation
(2)	Separable investment	65000	85000	45000	195000	Calculate the scales and investments of any two functions. The separable investment is the difference between the fixed-asset investment and the difference in investment between the two functions
(3)	Remaining investment				55000	(1) − (2)

Continued

Serial number	Item	Flood control	Water supply	Power generation	Total	Note
(4)	Annual benefit of different function projects	112000	145000	68000	325000	
(5)	Annual costs of alternative project	180000	176000	85000	441000	
(6)	Appropriate cost	112000	145000	68000	325000	The smaller one of NO. 4 and NO. 5
(7)	Separable costs	78000	92000	48000	218000	Difference between separable investment and the operation cost of this function project
(8)	Remaining benefits	34000	53000	20000	107000	(6) − (7)
(9)	Proportions in remaining benefits of different functions/%	31.78	49.53	18.69	100.00	(8) ÷ 107000
(10)	Allocation of remaining investment	17477	27243	10280	55000	(9) × (3)
(11)	The investment in sharing project	82477	112243	55280	250000	(2) + (10)

6. Allocation according to the primary and secondary level of each function

For multi-purpose water conservancy project, a certain function is dominant, the operation mode of the project is required to comply with its requirements. On the other hand, the amount and time of water supply of other secondary functions are subordinate. In this case, if the benefit of the main function accounts for a large proportion, the project can bear most of the cost, while the secondary function only bears the separable cost or special project cost.

Due to the complexity of investment and cost allocation of multi-purpose project, it is difficult to find a widely-used allocation method for all kinds of multi-purpose water conservancy construction projects. For the investment allocation of key projects, apply multiple methods and select a reasonable allocation result or the averaged value of the methods.

For the total cost, the allocation method is the same as that of the investment. To simplify the calculation, the same proportion as the investment allocation can be adopted. However, for the cost of a certain function, such as the water resource fee for urban water supply, power generation, etc., it should be treated as a special cost.

8.5 Rationality Analysis of Cost Allocation

Cost allocation of multi-purpose water conservancy project is complex as many factors are involved. Therefore, in practice, it is necessary to analyze the rationality of the results of cost allocation.

(1) The allocated cost shall not be greater than the corresponding benefit.

(2) The allocated cost shall not be greater than that of the optimal equivalent alternative.

(3) The allocated cost shall not be greater than that of the project constructed separately.

(4) The sum of the allocated cost shall be equal to the total cost of the project.

9

Case Study of Economic Evaluation of Hydropower Projects

9.1 Project Overview

Q Hydropower Station is in ×× Province. The main task of the project is hydropower generation. Adopts the damming and diversion development, the project mainly composed of water retaining structure, water release structure, diversion tunnel, powerhouse, etc. It is a daily regulating reservoir. The normal storage water level and dead water level of the reservoir is 2743m and 2739m, respectively. The total storage capacity is 1.73 million m^3. The installed capacity of the power station is 210 MW, including the working and duplicate capacity as 160 MW, 50 MW, respectively. The annual average energy output is 697.3 million kW·h.

The construction period of the project is 5 years, and the construction investment is 1640.24 million yuan.

Assume that the annual average energy output, since the 6th year, is 697.3 million kW·h. This energy value deducting the auxiliary power consumption and line loss is the on-grid energy. The effective power utilization coefficient is 0.965. The auxiliary power consumption rate is 0.5%. Herein, the lines loss is not considered. The investment process and unit operation plan are provided in Table 9.1.

Table 9.1 Investment process and unit operation plan

Serial number	Item	1	2	3	4	5	Six years later	Total
1	Construction investment/10^4 yuan	23607	33773	36550	43286	26809		164024
2	Installed capacity at the end of year/MW					210	210	
3	Annual power generation/(10^8 kW·h)						6.973	
4	Effective power/(10^8 kW·h)						6.729	

9.3 National Economic Evaluation

Continued

Serial number	Item	1	2	3	4	5	Six years later	Total
5	Plant power/(10^8 kW·h)						6.695	
6	On-grid energy/(10^8 kW·h)						6.695	

9.2 Purpose and Basis of Economic Evaluation

9.2.1 Evaluation Purpose

In accordance with the relevant laws, regulations and current financial system of the government, the economic evaluation of the project aims at the two following aspects: Evaluate the economic feasibility of the project, including the analysis of the economic benefits and costs of the project, and calculate the economic indicators of the project. On the other hand, evaluate the economic and financial feasibility of the project, including the calculation of the project cost, the analysis of the project profit ability, debt-paying ability, and survivability.

9.2.2 Evaluation Basis

(1) *Construction Projects Economic Evaluation Method and Parameter* (*3rd Edition*) (*fgtz* [2006] *No. 1325*).

(2) *Specification on Economic Evaluation of Hydropower Project* (DL/T 5441-2010).

9.3 National Economic Evaluation

9.3.1 Engineering Cost

1. Construction investment

The investment in the national economic evaluation is the part of the budget estimate that provides the investment minus the transfer payment. For this project, the construction investment is 1640.24 million yuan at the market price. Therefore, the national economic evaluation does not need to adjust the price, but the taxes, as belonging to the internal transfer payment of the national economy, are excluded. Specifically, the taxes include the construction tax, the land occupation tax, the land reclamation fee and the forest vegetation restoration fee. For this project, the total tax is 52.8 million yuan. Specifically, the construction tax is 51.45 million yuan, the estimated land occupation tax, land reclamation fee and forest vegetation restoration fee are 1.35 million yuan. Therefore, the construction investment after deducting the taxes is 1587.44 million yuan.

2. Operating cost

The operation cost is estimated using the production factor cost method. Herein, the amortization of intangible assets and other assets are not considered. The production cost includes the material cost, wage and welfare cost, repair cost, insurance cost, reservoir fund, and other expenses, etc. In particular, the water resources fee is excluded as it is the internal transfer of national economy. Details are referred to Section 9.4.1. The annual operation cost is 35.35 million yuan.

3. Working capital

Working capital is the capital occupied and used for a long time during the operation period. For this project, the estimated working capital is 2.1 million yuan given that 10 yuan/kW of the installed capacity. It is invested in the 5th year.

9.3.2 Engineering Benefit

The power generation benefit of Q Hydropower Station is calculated by the cost of thermal power station, which is the best equivalent alternative power station.

According to the results of power balance, the working capacity of the power station in the power grid is 160 MW, and the annual generating capacity is 697.3 million kW·h. The capacity coefficient of the alternative thermal power station is 1.1. The electricity coefficient is 1.05. The installed capacity of the alternative thermal power station is 176 MW. The annual generating capacity of the alternative thermal power station is 732.2 million kW·h.

According to data of the new thermal power station, the construction investment is 4500 yuan/kW, the fixed operation rate is 4%, the coal consumption rate is 350 g/(kW·h), and the standard coal price is 400 yuan/t. Therefore, the main parameters of the alternative thermal power station are as follows:

investment: 4500 yuan/kW;

coal consumption: 350g/(kW·h);

standard coal price: 400 yuan/t;

fixed operating rate: 4%.

Accordingly, the construction investment, the fuel cost and the operating cost is 792 million yuan, 102.5 million yuan, and 31.68 million yuan, respectively.

The construction investment is invested in three years at the rates of 30%, 40% and 30% respectively. Renovation cost is needed after 25 years operation of the thermal power station. It is estimated as 75% of the construction investment and invested in two years.

9.3.3 Methods and Results of National Economic Evaluation

According to *Construction Projects Economic Evaluation Method and Parameter (3rd Edition)*, the social discount rate of the project is 8%.

The construction and production period of the project is 5 years and 50 years, respectively. Accordingly, the economic calculation period is 55 years. The first construction year is assumed as the base year, and its beginning is assumed as the conversion starting point.

According to *Construction Projects Economic Evaluation Method and Parameter (3rd Edition)*, the national economic evaluation indicators include economic net present value (ENPV), economic internal rate of return (EIRR), benefit cost ratio (RBC), etc.

The economic service life of the mechanical and electrical equipment of the hydropower station is 25 years. When it is due, the equipment shall be renewed according to the original scale. The renewal cost shall be invested in two years at the rates of 50% and 50%, respectively.

See Table 9.2 for the economic benefit cost flow.

The net benefit flow is calculated by using Excel-internal function IRR (value) and the value of IRR is 9.65%. The net benefit flow is calculated by using Excel-internal function NPV, and the value of ENPV is 170.09 million. The alternative cost (benefit) flow is calculated by using Excel-internal function NPV, and the present value of benefit is 1757.22 million yuan. The present value of cost is 1587.13 million yuan calculated by using the Excel-internal function NPV. And the value of EBCR (present value of benefit/present value of cost) is 1.11.

Accordingly, for the Q Hydropower Station: the EIRR is 9.65%, larger than the social discount rate of 8%. ENPV is 170.09 million yuan, larger than zero. RBC is 1.11, larger than 1.0. These indicate that the project is economic and feasible.

9.3.4 Sensitivity Analysis

Sensitivity analysis is a kind of uncertainty analysis methods. Firstly, it finds out the sensitive factors of the economic benefit index of the investment project. Then, the degree of impacts and sensitivity of the identified factors are analyzed and calculated. Finally, the risk tolerance of the project is evaluated. In this project, the main sensitive factors are construction investment and valid energy. Single-factor sensitivity analysis is conducted here, and four schemes are designed as follows:

(1) The construction investment increased by 10%, the valid energy unchanged.
(2) The construction investment reduced by 10%, the valid energy unchanged.
(3) The valid energy increased by 10%, the construction investment unchanged.
(4) The valid energy reduced by 10%, the construction investment unchanged.

9 Case Study of Economic Evaluation of Hydropower Projects

Table 9.2 Economic benefit – cost statement

unit: 10^4 yuan

Serial number	Item	Total	Construction period					Operation period											
			1	2	3	4	5	6	7	8	9	10	11	12	13	14	15	16	17
	Installed capacity at the end of year/MW						210	210	210	210	210	210	210	210	210	210	210	210	210
	Annual effective capacity/MW						210	210	210	210	210	210	210	210	210	210	210	210	210
	Annual effective on-grid energy/(10^8 kW·h)							6.695	6.695	6.695	6.695	6.695	6.695	6.695	6.695	6.695	6.695	6.695	6.695
1	Cost (benefit) flow of alternative project	809516	0	0	23760	31680	23760	13418	13418	13418	13418	13418	13418	13418	13418	13418	13418	13418	13418
1.1	Construction investment	79200	0	0	23760	31680	23760	0	0	0	0	0	0	0	0	0	0	0	0
1.2	Maintain operation investment	59400	0	0	0	0	0	0	0	0	0	0	0	0	0	0	0	0	0
1.3	Operating cost	158400	0	0	0	0	0	3168	3168	3168	3168	3168	3168	3168	3168	3168	3168	3168	3168
1.4	Fuel cost	512516	0	0	0	0	0	10250	10250	10250	10250	10250	10250	10250	10250	10250	10250	10250	10250
1.5	Others	0	0	0	0	0	0	0	0	0	0	0	0	0	0	0	0	0	0
2	Cost flow of designed scheme	371536	22847	32686	35373	41893	26156	3535	3535	3535	3535	3535	3535	3535	3535	3535	3535	3535	3535
2.1	Construction investment	158744	22847	32686	35373	41893	25946	0	0	0	0	0	0	0	0	0	0	0	0
2.2	Maintain operation investment	35826	0	0	0	0	0	0	0	0	0	0	0	0	0	0	0	0	0
2.3	Operating cost	176756	0	0	0	0	0	3535	3535	3535	3535	3535	3535	3535	3535	3535	3535	3535	3535
2.4	Working capital	210	0	0	0	0	210	0	0	0	0	0	0	0	0	0	0	0	0
2.5	Project indirect cost	0	0	0	0	0	0	0	0	0	0	0	0	0	0	0	0	0	0
3	Net benefit flow	437980	−22847	−32686	−11613	−10213	−2396	9883	9883	9883	9883	9883	9883	9883	9883	9883	9883	9883	9883

9.3 National Economic Evaluation

Continued

Series number	Item	Operation period																		
		18	19	20	21	22	23	24	25	26	27	28	29	30	31	32	33	34	35	36
	Installed capacity at the end of year/MW	210	210	210	210	210	210	210	210	210	210	210	210	210	210	210	210	210	210	210
	Annual effective capacity/MW	210	210	210	210	210	210	210	210	210	210	210	210	210	210	210	210	210	210	210
	Annual effective on-grid energy/(10^8 kW·h)	6.695	6.695	6.695	6.695	6.695	6.695	6.695	6.695	6.695	6.695	6.695	6.695	6.695	6.695	6.695	6.695	6.695	6.695	6.695
1	Cost (benefit) flow of alternative project	13418	13418	13418	13418	13418	13418	13418	13418	13418	13418	13418	13418	43118	43118	13418	13418	13418	13418	13418
1.1	Construction investment	0	0	0	0	0	0	0	0	0	0	0	0	0	0	0	0	0	0	0
1.2	Maintain operation investment	0	0	0	0	0	0	0	0	0	0	0	0	29700	29700	0	0	0	0	0
1.3	Operating cost	3168	3168	3168	3168	3168	3168	3168	3168	3168	3168	3168	3168	3168	3168	3168	3168	3168	3168	3168
1.4	Fuel cost	10250	10250	10250	10250	10250	10250	10250	10250	10250	10250	10250	10250	10250	10250	10250	10250	10250	10250	10250
1.5	Others	0	0	0	0	0	0	0	0	0	0	0	0	0	0	0	0	0	0	0
2	Cost flow of designed scheme	3535	3535	3535	3535	3535	3535	3535	3535	3535	3535	3535	3535	21448	21448	3535	3535	3535	3535	3535
2.1	Construction investment	0	0	0	0	0	0	0	0	0	0	0	0	0	0	0	0	0	0	0
2.2	Maintain operation investment	0	0	0	0	0	0	0	0	0	0	0	0	17913	17913	0	0	0	0	0
2.3	Operating cost	3535	3535	3535	3535	3535	3535	3535	3535	3535	3535	3535	3535	3535	3535	3535	3535	3535	3535	3535
2.4	Working capital	0	0	0	0	0	0	0	0	0	0	0	0	0	0	0	0	0	0	0
2.5	Project indirect cost	0	0	0	0	0	0	0	0	0	0	0	0	0	0	0	0	0	0	0
3	Net benefit flow	9883	9883	9883	9883	9883	9883	9883	9883	9883	9883	9883	9883	21670	21670	9883	9883	9883	9883	9883

Continued

Series number	Item	Operating cost																		
		37	38	39	40	41	42	43	44	45	46	47	48	49	50	51	52	53	54	55
	Installed capacity at the end of year/MW	210	210	210	210	210	210	210	210	210	210	210	210	210	210	210	210	210	210	210
	Annual effective capacity/MW	210	210	210	210	210	210	210	210	210	210	210	210	210	210	210	210	210	210	210
	Annual effective on - grid energy/(10^8 kW • h)	6.695	6.695	6.695	6.695	6.695	6.695	6.695	6.695	6.695	6.695	6.695	6.695	6.695	6.695	6.695	6.695	6.695	6.695	6.695
1	Cost (benefit) flow of alternative project	13418	13418	13418	13418	13418	13418	13418	13418	13418	13418	13418	13418	13418	13418	13418	13418	13418	13418	13418
1.1	Construction investment	0	0	0	0	0	0	0	0	0	0	0	0	0	0	0	0	0	0	0
1.2	Maintain operation investment	0	0	0	0	0	0	0	0	0	0	0	0	0	0	0	0	0	0	0
1.3	Operation cost	3168	3168	3168	3168	3168	3168	3168	3168	3168	3168	3168	3168	3168	3168	3168	3168	3168	3168	3168
1.4	Fuel cost	10250	10250	10250	10250	10250	10250	10250	10250	10250	10250	10250	10250	10250	10250	10250	10250	10250	10250	10250
1.5	Others	0	0	0	0	0	0	0	0	0	0	0	0	0	0	0	0	0	0	0
2	Cost flow of designed scheme	3535	3535	3535	3535	3535	3535	3535	3535	3535	3535	3535	3535	3535	3535	3535	3535	3535	3535	3535
2.1	Construction investment	0	0	0	0	0	0	0	0	0	0	0	0	0	0	0	0	0	0	0
2.2	Maintain operation investment	0	0	0	0	0	0	0	0	0	0	0	0	0	0	0	0	0	0	0
2.3	Operating cost	3535	3535	3535	3535	3535	3535	3535	3535	3535	3535	3535	3535	3535	3535	3535	3535	3535	3535	3535
2.4	Working capital	0	0	0	0	0	0	0	0	0	0	0	0	0	0	0	0	0	0	0
2.5	Project indirect cost	0	0	0	0	0	0	0	0	0	0	0	0	0	0	0	0	0	0	0
3	Net benefit flow	9883	9883	9883	9883	9883	9883	9883	9883	9883	9883	9883	9883	9883	9883	9883	9883	9883	9883	9883

Calculation indicators: FIRR 9.65%; ENPV ($i_s=8\%$) 17009 10^4 yuan; EBCR ($i_s=8\%$) 1.11.

9.4 Financial Evaluation

The sensitivity analysis results are provided in Table 9.3.

Table 9.3   Results of sensitivity analysis

Variation Range	FIRR/%	ENPV/10⁴ yuan	Economic benefit cost ratio, EBCR
Schemes	9.65	17009	1.11
Construction investment increased 10%, effective energy unchanged	10.36	25581	1.15
Construction investment decreased 10%, effective energy unchanged	14.16	50997	1.36
Effective energy increased 10%, construction investment unchanged	12.91	46818	1.32
Effective energy decreased 10%, construction investment unchanged	9.16	11673	1.08

Results of sensitivity analysis show that when the investment or benefit is changing toward the unfavorable direction, the economic internal rate of return (EIRR) is higher than 8%. This indicates that the project has some anti-risk ability.

9.4 Financial Evaluation

9.4.1 Engineering Cost

For hydropower projects, the main financial expenses include total investment, cost expense and taxes.

1. Total investment

The total investment consists of construction investment, interest during construction and working capital.

The total investment of Q Hydropower Station project is 1828.54 million yuan, including 1640.24 million yuan of construction investment, 186.2 million yuan of interest during the construction period and 2.1 million yuan of working capital.

The capital of the project accounts for 20% of the total investment, and the rest is borrowed from the bank. The term of the loan is 20 years, of which the grace period is 5 years. The annual interest rate of domestic banks' loans over five years is 5.65%. The interest during the construction period is calculated by compound interest method and included in the original value of the fixed assets of the project.

The working capital is estimated by expanding index method, which is 2.1 million yuan based on 10 yuan/kW. 30% of the working capital is self-raised, 70% is borrowed from the bank, and the annual interest rate of the loan is 5.1%.

The total investment plan and fund raising of the project is provided in Table 9.4.

9 Case Study of Economic Evaluation of Hydropower Projects

Table 9.4   Total investment and fund raising of the project   unit: 10^4 yuan

Serial number	Item	Total	Construction period				
			1	2	3	4	5
1	Total investment	182854	24137	35617	40057	48745	34299
1.1	Construction investment	164024	23607	33773	36550	43286	26809
1.2	Interest during construction period	18620	531	1844	3507	5459	7280
1.3	Working capital	210	0	0	0	0	210
2	Fund raising	182854	24137	35617	40057	48745	34299
2.1	Project capital fund	36571	4827	7123	8011	9749	6860
2.1.1	For construction investment	36508	4827	7123	8011	9749	6797
2.1.2	For working capital	63	0	0	0	0	63
2.2	Debt fund	146284	19310	28493	32046	38996	27439
2.2.1	For construction investment	127516	18779	26649	28538	33537	20012
2.2.2	For interest during construction period	18620	531	1844	3507	5459	7280
2.2.3	For working capital	147	0	0	0	0	147

2. Total cost expense

The total cost expense is all the expenses to produce power products or the provision of various power auxiliary services during the hydropower operation. It consists of operating costs, depreciation expenses, amortization expenses and interest expenses. Operating costs include material cost, wage and welfare cost, repair cost, water resources fee, insurance cost, reservoir fund, other expenses, etc.

(1) Material cost. Material cost refers to the cost of raw materials, raw water, auxiliary materials, spare parts, etc. consumed in the operation and maintenance of water conservancy and hydropower projects. As there is no data, this time the value is taken according to the *Interim Provisions on financial evaluation of hydropower construction projects* (for Trial Implementation). The installed capacity of the power station is 210 MW, and the material cost is 5 yuan/kW, which is 1.05 million yuan.

(2) Salary and welfare cost. Salary and welfare cost refers to various forms of remuneration and other related expenses given to obtain the services provided by the employees, including the employees' wages (including wages, bonuses, allowances, subsidies and other monetary remuneration), employee welfare expenses, labor union funds, employee education funds, housing fund, medical insurance premiums, endowment insurance premiums, unemployment insurance premiums, work-related injury insurance premiums and maternity insurance premium and other basic social insurance premium.

According to the salary of urban residents in the project area in the past three years, the annual average salary of employees is 50000 yuan, and the total annual salary of employees is 2.1 million yuan. The employee welfare expenses are estimated at 62% of the total wages of the employees, including 14% of the employee welfare expenses, 9% of the medical insurance expenses, 20% of the endowment insurance expenses, 2% of the unemployment insurance expenses, 1.5% of the industrial injury insurance expenses, 1% of the maternity insurance expenses, 10% of the housing fund, 2% of the labor union expenses and 2.5% of the employee education expenses. The annual employee compensation is 3.4 million yuan.

(3) Repair cost. Repair cost refers to the cost of necessary repair to maintain the normal operation and use of fixed assets and give full play to the use efficiency. According to the size of repair scope and repair interval, it can be divided into major repair and minor repair. Considering the characteristics of the project, the annual repair cost of the project is estimated as 1% of the original value of fixed assets, and the annual repair cost is 18.26 million yuan.

(4) Water resources fee. According to the local documents of the project area, the water resource fee for hydropower generation is charged as 0.003 yuan/(kW·h), the annual power generation of the project is 697.3 million kW·h, and the annual water resource fee is 2.09 million yuan.

(5) Premium. The insurance premium includes fixed assets insurance and other insurance. It is estimated to be 0.25% of the original value of fixed assets, and the annual insurance premium is 4.57 million yuan.

(6) Reservoir fund. The reservoir fund refers to the expenses needed to support the implementation of the infrastructure construction and economic development planning of the reservoir area and the resettlement area, support the reservoir area protection project and the production and maintenance of living facilities, and solve other remaining problems of the reservoir immigrants after the reservoir impoundment.

According to the *Interim Measures for the administration of the collection and use of funds for large and medium-sized reservoir areas formulated by the Ministry of finance*, the state integrates the funds for the maintenance of the original reservoir area, the funds for the later stage support of the original reservoir area and the funds for the later stage support of the resettlement undertaken by the operational large and medium-sized reservoirs (referring to the reservoirs and hydropower stations with an installed capacity of 25000kW and above and generating revenue), and establishes the funds for the large and medium-sized reservoir area (referred to as the "reservoir area fund"), which is mainly used to support the implementation of infrastructure construction and economic development planning in the reservoir area and resettlement area, the reservoir area protection project and the maintenance of production and living facilities of the immigrants, and to solve other remaining problems of the reservoir immigrants. The reservoir fund is raised from the power generation income of large and medium-sized reservoirs with power generation income, and is collected according to the actual online sales power of the reservoir and the standard of 0.008 yuan/(kW·h). The fund in the reservoir area of the project is charged at 0.008 yuan/(kW·h). The annual power generation of the project is 697.3 million kW·h, the on-grid power is 669.5 million kW·h, and the annual reservoir fund is 5.36 million yuan.

(7) Other expenses. Other expenses refer to the expenses directly related to the operation and maintenance of water conservancy and hydropower projects, except for the salaries of employees and materials. As there is no data, this time the value is taken according to the Interim Provisions on financial evaluation of hydropower construction projects (Trial), the installed capacity of the power station is 210MW, other costs are estimated at 24 yuan/kW, and the annual other costs are 5.04 million yuan.

(8) Depreciation. The depreciation cost is the depreciation cost of fixed assets. The fixed assets of the project will be worn out in the use process, and the loss of value is usually compensated by the way of drawing depreciation. Straight line method is adopted for depreciation of the project. The residual value is 0, the comprehensive depreciation life is 30 years, and the average annual comprehensive depreciation rate is 3.3%.

$$\text{Depreciation cost} = (\text{value of fixed assets} - \text{residual value of the project}) \times \text{comprehensive depreciation rate}$$

The annual depreciation cost of the project is 60.27 million yuan.

(9) Amortization fee. Amortization of intangible assets and other assets is not considered in the project, so the amortization fee is 0.

(10) Interest expense. Interest expense includes long-term loan interest, short-term loan interest and current loan interest. Long-term loan interest is calculated based on compound interest. Interest in construction period is included in the original value of fixed assets of the project, and that in operation period is included in production cost.

(11) Operating costs. The operation cost of **Q** Hydropower Station includes material cost, salary and welfare cost, repair cost, water resource cost, insurance cost, reservoir fund and other expenses. The total annual operation cost is 39.77 million yuan.

3. Taxes

Electricity sales tax includes value-added tax, sales tax surcharges and income tax.

According to the *Provisional Regulations of the People's Republic of China on Value Added Tax* implemented on January 1, 2009, the output tax of electric power products = sales amount × tax rate, the tax rate of value-added tax for hydropower stations greater than or equal to 50MW is 17%, and the tax rate of value-added tax for hydropower stations less than 50MW is 6%. Value added tax is based on sales revenue and the tax rate is 17%. Value added tax is non-price tax and only serves as the basis for the calculation of sales tax surcharges.

Sales tax surcharges include urban maintenance and construction tax, education surcharges and local education surcharges, which belong to the intra price tax.

9.4 Financial Evaluation

According to the *Interim Regulations of the People's Republic of China on Urban Maintenance and Construction Tax (GF [1985] No. 19)*, in order to strengthen the maintenance and construction of cities, expand and stabilize the sources of urban maintenance and construction funds, and levy urban maintenance and construction tax, the urban maintenance and construction tax is calculated on the basis of the product tax, value-added tax and business tax actually paid by the taxpayers, respectively with the product tax, value-added tax Business tax shall be paid at the same time. The urban maintenance and construction tax rate are as follows: if the taxpayer is in the urban area, the tax rate is 7%; if the taxpayer is located in the county or town, the tax rate is 5%; if the taxpayer is not located in the urban area, county or town, the tax rate is 1%.

According to the *Interim Provisions on the Collection of Educational Surcharges* (issued by the State Council on April 28, 1986), in order to implement the decision of the Central Committee of the Communist Party of China on the reform of education system, accelerate the development of local education, expand the source of funds for local education funds, collect Educational Surcharges, which are based on the value-added tax, business tax and consumption tax actually paid by various units and individuals for the purpose of tax calculation, the additional rate of education fee is 3%, which shall be paid together with value-added tax, business tax and consumption tax.

According to the *Circular of the Ministry of Finance on Issues Related to the Unification of Local Education Additional Policies (CZ [2010] No. 98)*, in order to implement the outline of the *National Medium -and Long -term Education Reform and Development Plan (2010 - 2020)*, further standardize and expand the financing channels of financial education funds, support the development of local education, and unify the local education additional collection standards. The collection standard of local education surcharges shall be 2% of the value-added tax, business tax and consumption tax actually paid by units and individuals (including enterprises with foreign investment, foreign enterprises and foreign individuals). For provinces that have been approved by the Ministry of Finance and whose collection standard is less than 2%, the collection standard of local education surcharges should be adjusted to 2%.

The income tax is calculated based on the taxable income. The taxable income of the project is the balance of the sales revenue of power generation after deducting the cost and sales tax surcharges. According to the newly promulgated *Enterprise Income Tax Law of the People's Republic of China*, the income tax is levied at 25%. The new project of surface water source project is listed in the *Catalogue of Preferential Corporate Income Tax for Public Infrastructure Projects (2008 Edition)*. Since the first tax year of the project's production and operation income, corporate income tax will be exempted from the first year to the third year and reduced by half from the fourth year to the sixth year.

The income tax is based on the generation profit and the tax rate is 25%. According to several policies and measures of the State Council for the development of the western region, the enterprise income tax of newly established enterprises in the western region, such as transportation, electric power, water conservancy, postal service, radio, television, etc., shall be exempted for two years and reduced by half for three years. This electricity price calculation shall be carried out in accordance with this provision.

9.4.2 Generation Revenue and Profit Calculation

1. Generation revenue

$$\text{Generation revenue} = \text{on-grid energy} \times \text{on-grid price}$$

The on-grid energy is the effective power after deducting the auxiliary power consumption and the loss of power transmission and transformation.

2. Total profit

$$\text{Total profit} = \text{generation revenue} - \text{total cost} - \text{sales tax surcharges}$$

$$\text{After tax profit} = \text{total profit income tax}$$

After the after-tax profit is extracted from 10% of the statutory surplus reserve, the remaining part is the distributable profit, and then the payable profit distributed to the investors is the undistributed profit.

9.4.3 On-grid Price Estimation

In this financial evaluation, the calculation period is 35 years. Specifically, the calculation time during the construction period is 5 years and another 30 years is during the operation period.

The loan repayment period is 25 years, and the grace period is 5 years. The repayment method is equal principal and interest. The repayment fund is all undistributed profits and depreciation expenses.

According to the requirements of project owners and FIRR of the project capital of 8%, the estimated on-grid price is 0.296 yuan/(kW·h). According to field survey, for the power grid of Q Hydropower Station, the on-grid price of is 0.251 - 0.3 yuan/(kW·h) (excluding VAT). Comparatively, the on-grid price of this project, estimated as 0.296 yuan/(kW·h), is locally competitive.

Financial indicators are listed in Table 9.5. Other financial statements are provided in Tables 9.6 - Tables 9.12.

Table 9.5 Financial evaluation indicators

Serial number	Item	Unit	Value
1	Total investment	10^4 yuan	182854
1.1	Construction investment	10^4 yuan	164024
1.2	Construction period interest	10^4 yuan	18620

Continued

Serial number	Item	Unit	Value
1.3	Working capital	10⁴ yuan	210
2	On-grid energy during operation period	yuan/(kW·h)	0.296
3	Total sales revenue of power generation	10⁴ yuan	594543
4	Total sales tax surcharges	10⁴ yuan	10107
5	Total generation profit	10⁴ yuan	180766
6	Profitability index		
6.1	Profitability indicators	%	5.87
6.2	FIRR of capital fund	%	8
6.3	Payback period		17.5
7	Repayment ability index		
7.1	Payback period of loan		25
7.2	Asset liability ratio (5th year, maximum value)	%	80

9.4.4 Financial Evaluation

1. Financial viability analysis

According to the cash-flow statement of the financial plan as well as the recommended on-grid price of 0.296 yuan/(kW·h), the accumulated surplus cash in each year is all larger than 0. It indicates that the project has financial viability to maintain normal operation.

2. Solvency analysis

According to the loan repayment statement, the project loan is paid in the way of equal principal and interest, and the funds used for repayment include undistributed profits and depreciation expenses. The recommended on-grid price is 0.296 yuan/(kW·h). The interest reserve ratio is larger than 1.1 during the loan repayment period, and it would be larger than 2 after the 18th year. The debt service reserve ratio is larger than 1 during the loan repayment period, indicating that the project has strong debt service ability.

According to the balance sheet, the maximum asset liability ratio of this project is 80%. As time goes by with the loan repayment, the asset liability ratio of the power station at the end is 0.1%.

3. Profitability analysis

According to the cash flow statements of the project investment and project capital, when the recommended on-grid price is 0.296 yuan/(kW·h), the FIRR of all investment after income tax is 5.87%, and the financial internal rate of return of capital is 8.0%. The investment payback period is 17.5 years, and all investment can be recovered in the 11th year after all units are put into operation. It shows that the profitability of the project is strong.

9 Case Study of Economic Evaluation of Hydropower Projects

Table 9.6   Schedule of loan repayment

unit: 10^4 yuan

Serial number	Item	Total	Construction period					Operation period																	
			1	2	3	4	5	6	7	8	9	10	11	12	13	14	15	16	17	18	19	20	21	22	23	24	25
1	Loan																									
1.1	Beginning balance	1994218	0	19310	47803	79849	118845	146137	142012	137654	133051	128187	123048	117619	111883	105823	99421	92657	85511	77961	69985	61558	52655	43248	33311	22811	11719
1.2	Repayment of principal and interest at current period	210481	0	0	0	0	0	12381	12381	12381	12381	12381	12381	12381	12381	12381	12381	12381	12381	12381	12381	12381	12381	12381	12381	12381	12381
	Including repayment of principal	112826	0	0	0	0	0	4125	4358	4604	4864	5139	5429	5736	6060	6402	6764	7146	7550	7976	8427	8903	9406	9938	10499	11092	11719
	Interest payment	101488	0	0	0	0	0	8257	8024	7777	7517	7243	6952	6645	6321	5979	5617	5235	4831	4405	3954	3478	2975	2444	1882	1289	662
1.3	Loan balance at the end of the period	2027528	19310	47803	79849	118845	146137	142012	137654	133051	128187	123048	117619	111883	105823	99421	92657	85511	77961	69985	61558	52655	43248	33311	22811	11719	0
2	Short term loan																										
2.1	Debt balance at the start of the period	0	0	0	0	0	0	0	0	0	0	0	0	0	0	0	0	0	0	0	0	0	0	0	0	0	0
2.2	Repayment of principal and interest at current period	0	0	0	0	0	0	0	0	0	0	0	0	0	0	0	0	0	0	0	0	0	0	0	0	0	0
	Including repayment of principal	0	0	0	0	0	0	0	0	0	0	0	0	0	0	0	0	0	0	0	0	0	0	0	0	0	0
	Interest payment	0	0	0	0	0	0	0	0	0	0	0	0	0	0	0	0	0	0	0	0	0	0	0	0	0	0
2.3	Debt balance at the end of the period	0	0	0	0	0	0	0	0	0	0	0	0	0	0	0	0	0	0	0	0	0	0	0	0	0	0
3	Total loans																										
3.1	Beginning balance	1994218	0	19310	47803	79849	118845	146137	142012	137654	133051	128187	123048	117619	111883	105823	99421	92657	85511	77961	69985	61558	52655	43248	33311	22811	11719
3.2	Repayment of principal and interest at current period	210481	0	0	0	0	0	12381	12381	12381	12381	12381	12381	12381	12381	12381	12381	12381	12381	12381	12381	12381	12381	12381	12381	12381	12381
	Including repayment of principal	112826	0	0	0	0	0	4125	4358	4604	4864	5139	5429	5736	6060	6402	6764	7146	7550	7976	8427	8903	9406	9938	10499	11092	11719

9.4 Financial Evaluation

Continued

Serial number	Item	Total	Construction period					Operation period																	
			1	2	3	4	5	6	7	8	9	10	11	12	13	14	15	16	17	18	19	20	21	22	23	24	25
	Interest payment	97655	0	0	0	0	0	8257	8024	7777	7517	7243	6952	6645	6321	5979	5617	5235	4831	4405	3954	3478	2975	2444	1882	1289	662
3.3	Remaining balance at the end of the period	2027528	19310	47803	79849	118845	146137	142012	137654	133051	128187	123048	117619	111883	105823	99421	92657	85511	77961	69985	61558	52655	43248	33311	22811	11719	0
	Calculation indicators																										
	Interest coverage ratio							1.15	1.18	1.22	1.26	1.31	1.36	1.43	1.50	1.59	1.69	1.81	1.96	2.15	2.40	2.72	3.19	3.88	5.04	7.35	14.31
	Debt service coverage ratio							1.25	1.25	1.24	1.23	1.23	1.20	1.20	1.19	1.18	1.17	1.17	1.16	1.15	1.14	1.13	1.12	1.11	1.10	1.09	1.07

Table 9.7 Total cost estimation

unit: 10^4 yuan

Serial number	Item	Total	Construction period					Operation period												
			1	2	3	4	5	6	7	8	9	10	11	12	13	14	15	16	17	18
1	Material fee	3150	0	0	0	0	0	105	105	105	105	105	105	105	105	105	105	105	105	105
2	Salary and welfare expense	10206	0	0	0	0	0	340	340	340	340	340	340	340	340	340	340	340	340	340
3	Repair fee	54793	0	0	0	0	0	1826	1826	1826	1826	1826	1826	1826	1826	1826	1826	1826	1826	1826
4	Water resources fee	6276	0	0	0	0	0	209	209	209	209	209	209	209	209	209	209	209	209	209
5	Insurance premium	13698	0	0	0	0	0	457	457	457	457	457	457	457	457	457	457	457	457	457
6	Reservoir fund	16069	0	0	0	0	0	536	536	536	536	536	536	536	536	536	536	536	536	536
7	Other costs	15120	0	0	0	0	0	504	504	504	504	504	504	504	504	504	504	504	504	504
8	Operation cost	119312	0	0	0	0	0	3977	3977	3977	3977	3977	3977	3977	3977	3977	3977	3977	3977	3977
9	Depreciation cost	182644	0	0	0	0	0	6027	6027	6027	6027	6027	6027	6027	6027	6027	6027	6027	6027	6027
10	Amortization cost	0	0	0	0	0	0	0	0	0	0	0	0	0	0	0	0	0	0	0
11	Interest expenditure	101713	0	0	0	0	0	8264	8031	7785	7525	7250	6960	6653	6329	5987	5625	5243	4839	4412
11.1	Long-term loan interest expenditure	101488	0	0	0	0	0	8257	8024	7777	7517	7243	6952	6645	6321	5979	5617	5235	4831	4405

9 Case Study of Economic Evaluation of Hydropower Projects

Continued

Serial number	Item	Total	Construction period										Operation period								
			1	2	3	4	5	6	7	8	9	10	11	12	13	14	15	16	17	18
11.2	Loan interest expenditure of working capital	225	0	0	0	0	0	7	7	7	7	7	7	7	7	7	7	7	7	7
11.3	Other short-term loan interest expenditure	0	0	0	0	0	0	0	0	0	0	0	0	0	0	0	0	0	0
12	Total cost	403670	0	0	0	0	0	18269	18036	17789	17529	17254	16964	16657	16333	15991	15629	15247	14843	14417

Serial number	Item		Operation period																
		19	20	21	22	23	24	25	26	27	28	29	30	31	32	33	34	35
1	Material fee	105	105	105	105	105	105	105	105	105	105	105	105	105	105	105	105	105
2	Salary and welfare expense	340	340	340	340	340	340	340	340	340	340	340	340	340	340	340	340	340
3	Repair cost	1826	1826	1826	1826	1826	1826	1826	1826	1826	1826	1826	1826	1826	1826	1826	1826	1826
4	Water resources fee	209	209	209	209	209	209	209	209	209	209	209	209	209	209	209	209	209
5	Insurance premium	457	457	457	457	457	457	457	457	457	457	457	457	457	457	457	457	457
6	Reservoir fund	536	536	536	536	536	536	536	536	536	536	536	536	536	536	536	536	536
7	Other cost	504	504	504	504	504	504	504	504	504	504	504	504	504	504	504	504	504
8	Operating cost	3977	3977	3977	3977	3977	3977	3977	3977	3977	3977	3977	3977	3977	3977	3977	3977	3977
9	Depreciation cost	6027	6027	6027	6027	6027	6027	6027	6027	6027	6027	6027	6027	6027	6027	6027	6027	7854
10	Amortization cost	0	0	0	0	0	0	0	0	0	0	0	0	0	0	0	0	0
11	Interest expenditure	3962	3486	2982	2451	1890	1296	670	7	7	7	7	7	7	7	7	7	7
11.1	Long-term loan interest expenditure	3954	3478	2975	2444	1882	1289	662	0	0	0	0	0	0	0	0	0	0
11.2	Loan interest expenditure of working capital	7	7	7	7	7	7	7	7	7	7	7	7	7	7	7	7	7
11.3	Other short-term loan interest expenditure	0	0	0	0	0	0	0	0	0	0	0	0	0	0	0	0	0
12	Total cost	13966	13490	12987	12455	11894	11301	10674	10012	10012	10012	10012	10012	10012	10012	10012	10012	11838

9.4 Financial Evaluation

Table 9.8 Profit and profit distribution

Serial number	Item	Total	Construction period					Operation period											
			1	2	3	4	5	6	7	8	9	10	11	12	13	14	15	16	17	18
1	Sales income		0	0	0	0	0	19751	19751	19751	19751	19751	19751	19751	19751	19751	19751	19751	19751	19751
1.1	Effective on-grid energy/(10^8 kW·h)		0	0	0	0	0	6.695	6.695	6.695	6.695	6.695	6.695	6.695	6.695	6.695	6.695	6.695	6.695	6.695
1.2	Effective capacity/MW							210	210	210	210	210	210	210	210	210	210	210	210	210
1.3	On-grid energy/[yuan/(kW·h)]		0	0	0	0	0	0.296	0.296	0.296	0.296	0.296	0.296	0.296	0.296	0.296	0.296	0.296	0.296	0.296
1.4	Capacity price/[yuan/(kW·h)]		0	0	0	0	0	0	0	0	0	0	0	0	0	0	0	0	0	0
2	Deductible VAT amount/10^4 yuan	10107	0	0	0	0	0	0	0	0	0	0	0	0	0	0	0	0	0	0
3	Sales income and surcharges/10^4 yuan		0	0	0	0	0	337	337	337	337	337	337	337	337	337	337	337	337	337
4	Total cost/10^4 yuan	403670	0	0	0	0	0	18269	18036	17789	17529	17254	16964	16657	16333	15991	15629	15247	14843	14417
5	Subsidy income/10^4 yuan	0	0	0	0	0	0	0	0	0	0	0	0	0	0	0	0	0	0	0
6	Total profit/10^4 yuan	180766	0	0	0	0	0	1213	1446	1692	1952	2227	2517	2824	3148	3490	3852	4234	4638	5065
7	Make up for losses of previous years/10^4 yuan	0	0	0	0	0	0	0	0	0	0	0	0	0	0	0	0	0	0	0
8	Taxable income/10^4 yuan	180766	0	0	0	0	0	1213	1446	1692	1952	2227	2517	2824	3148	3490	3852	4234	4638	5065
9	Income tax/10^4 yuan	43793	0	0	0	0	0	0	0	211	244	278	629	706	787	873	963	1059	1159	1266
10	Net profit/10^4 yuan	136973	0	0	0	0	0	1213	1446	1480	1708	1948	1888	2118	2361	2618	2889	3176	3478	3798
11	Undistributed profit at the beginning of the period/10^4 yuan	0	0	0	0	0	0	0	0	0	0	0	0	0	0	0	0	0	0	0
12	Profit available for distribution/10^4 yuan	136973	0	0	0	0	0	1213	1446	1480	1708	1948	1888	2118	2361	2618	2889	3176	3478	3798
13	Withdrawal of statutory surplus reserve/10^4 yuan	13697	0	0	0	0	0	121	145	148	171	195	189	212	236	262	289	318	348	380
14	Profit available for distribution to investors/10^4 yuan	123275	0	0	0	0	0	1091	1301	1332	1537	1754	1699	1906	2125	2356	2600	2858	3131	3419
15	Profit distribution by investors/10^4 yuan	0	0	0	0	0	0	0	0	0	0	0	0	0	0	0	0	0	0	0
16	Undistributed profit/10^4 yuan	123275	0	0	0	0	0	1091	1301	1332	1537	1754	1699	1906	2125	2356	2600	2858	3131	3419
17	Profit before interest and tax/10^4 yuan	282479	0	0	0	0	0	9477	9477	9477	9477	9477	9477	9477	9477	9477	9477	9477	9477	9477
18	Profit before interest, tax, depreciation and amortization/10^4 yuan	465123	0	0	0	0	0	15504	15504	15504	15504	15504	15504	15504	15504	15504	15504	15504	15504	15504

9 Case Study of Economic Evaluation of Hydropower Projects

Continued

Serial number	Item	Operation period																
		19	20	21	22	23	24	25	26	27	28	29	30	31	32	33	34	35
1	Sales income	19751	19751	19751	19751	19751	19751	19751	19751	19751	19751	19751	19751	19751	19751	19751	19751	19751
1.1	Effective on-grid energy/(10^8 kW·h)	6.695	6.695	6.695	6.695	6.695	6.695	6.695	6.695	6.695	6.695	6.695	6.695	6.695	6.695	6.695	6.695	6.695
1.2	Effective capacity/MW	210	210	210	210	210	210	210	210	210	210	210	210	210	210	210	210	210
1.3	On-grid energy/[yuan/(kW·h)]	0.296	0.296	0.296	0.296	0.296	0.296	0.296	0.296	0.296	0.296	0.296	0.296	0.296	0.296	0.296	0.296	0.296
1.4	Capacity price/[yuan/(kW·h)]	0	0	0	0	0	0	0	0	0	0	0	0	0	0	0	0	0
2	Deductible VAT amount/10^4 yuan	0	0	0	0	0	0	0	0	0	0	0	0	0	0	0	0	0
3	Sales income and surcharges/10^4 yuan	337	337	337	337	337	337	337	337	337	337	337	337	337	337	337	337	337
4	Total cost/10^4 yuan	13966	13490	12987	12455	11894	11301	10674	10012	10012	10012	10012	10012	10012	10012	10012	10012	11838
5	Subsidy income/10^4 yuan	0	0	0	0	0	0	0	0	0	0	0	0	0	0	0	0	0
6	Total profit/10^4 yuan	5515	5991	6494	7026	7587	8181	8807	9469	9469	9469	9469	9469	9469	9469	9469	9469	7643
7	Make up for losses of previous years/10^4 yuan	0	0	0	0	0	0	0	0	0	0	0	0	0	0	0	0	0
8	Taxable income/10^4 yuan	5515	5991	6494	7026	7587	8181	8807	9469	9469	9469	9469	9469	9469	9469	9469	9469	7643
9	Income tax/10^4 yuan	1379	1498	1624	1756	1897	2045	2202	2367	2367	2367	2367	2367	2367	2367	2367	2367	1911
10	Net profit/10^4 yuan	4136	4493	4871	5269	5690	6135	6605	7102	7102	7102	7102	7102	7102	7102	7102	7102	5732
11	Undistributed profit at the beginning of the period/10^4 yuan	0	0	0	0	0	0	0	0	0	0	0	0	0	0	0	0	0
12	Profit available for distribution/10^4 yuan	4136	4493	4871	5269	5690	6135	6605	7102	7102	7102	7102	7102	7102	7102	7102	7102	5732
13	Withdrawal of statutory surplus reserve/10^4 yuan	414	449	487	527	569	614	661	710	710	710	710	710	710	710	710	710	573
14	Profit available for distribution to investors/10^4 yuan	3723	4044	4384	4742	5121	5522	5945	6392	6392	6392	6392	6392	6392	6392	6392	6392	5159
15	Profit distribution by investors/10^4 yuan	0	0	0	0	0	0	0	0	0	0	0	0	0	0	0	0	0
16	Undistributed profit/10^4 yuan	3723	4044	4384	4742	5121	5522	5945	6392	6392	6392	6392	6392	6392	6392	6392	6392	5159
17	Profit before interest and tax/10^4 yuan	9477	9477	9477	9477	9477	9477	9477	9477	9477	9477	9477	9477	9477	9477	9477	9477	7650
18	Profit before interest, tax, depreciation and amortization/10^4 yuan	15504	15504	15504	15504	15504	15504	15504	15504	15504	15504	15504	15504	15504	15504	15504	15504	15504

9.4 Financial Evaluation

Table 9.9　Financial plan of cash flow statement

unit: 10^4 yuan

Serial number	Item	Total	Construction period					Operation period												
			1	2	3	4	5	6	7	8	9	10	11	12	13	14	15	16	17	18
1	Net cash flow from operating activities	421330	0	−33773	−36550	−43286	−27019	15504	15504	15293	15260	15226	14875	14798	14717	14632	14541	14446	14345	14238
1.1	Cash inflow	594543	0	0	0	0	0	15504	15504	19818	19818	19818	19818	19818	19818	19818	19818	19818	19818	19818
1.1.1	Sales income	594543	0	0	0	0	0	19818	19818	19818	19818	19818	19818	19818	19818	19818	19818	19818	19818	19818
1.1.2	Substituted money on VAT	0	0	0	0	0	0	0	0	0	0	0	0	0	0	0	0	0	0	0
1.1.3	Subsidy income	0	0	0	0	0	0	0	0	0	0	0	0	0	0	0	0	0	0	0
1.1.4	Other inflows	0	0	0	0	0	0	0	0	0	0	0	0	0	0	0	0	0	0	0
1.2	Cash outflow	173212	0	0	0	0	0	4314	4314	4525	4558	4592	4943	5020	5101	5187	5277	5373	5473	5580
1.2.1	Operating cost	119312	0	0	0	0	0	3977	3977	3977	3977	3977	3977	3977	3977	3977	3977	3977	3977	3977
1.2.2	Input value added tax	0	0	0	0	0	0	0	0	0	0	0	0	0	0	0	0	0	0	0
1.2.3	Sales tax and surcharges	10107	0	0	0	0	0	337	337	337	337	337	337	337	337	337	337	337	337	337
1.2.4	Value added tax	0	0	0	0	0	0	0	0	0	0	0	0	0	0	0	0	0	0	0
1.2.5	Income tax	43793	0	0	0	0	0	0	0	211	244	278	629	706	787	873	963	1059	1159	1266
1.2.6	Other outflows	0	0	0	0	0	0	0	0	0	0	0	0	0	0	0	0	0	0	0
2	Net cash flow from investment activities	−164234	−23607	−33773	−36550	−43286	−27019	0	0	0	0	0	0	0	0	0	0	0	0	0
2.1	Cash inflow	0	0	0	0	0	0	0	0	0	0	0	0	0	0	0	0	0	0	0
2.2	Cash outflow	164234	23607	33773	36550	43286	27019	0	0	0	0	0	0	0	0	0	0	0	0	0
2.2.1	Construction investment	164024	23607	33773	36550	43286	26809	0	0	0	0	0	0	0	0	0	0	0	0	0
2.2.2	Maintain operational investment main	0	0	0	0	0	0	0	0	0	0	0	0	0	0	0	0	0	0	0

Continued

Serial number	Item	Total	Construction period					Operation period											
			1	2	3	4	5	6	7	8	9	10	11	12	13	14	15	16	17	18
2.2.3	Working capital	210	0	0	0	0	210	0	0	0	0	0	0	0	0	0	0	0	0	0
2.2.4	Other outflows	0	0	0	0	0	0	0	0	0	0	0	0	0	0	0	0	0	0
3	Net cash flow from financing activities	−83616	23607	33773	36550	43286	27019	—	—	—	—	—	—	—	—	—	—	—	—	—
3.1	Cash inflow	182854	24137	35617	40057	48745	34299	0	0	0	0	0	0	0	0	0	0	0	0	0
3.1.1	Project capital inflow	36571	4827	7123	8011	9749	6860	0	0	0	0	0	0	0	0	0	0	0	0	0
3.1.2	Construction investment loan	146137	19310	28493	32046	38996	27292	0	0	0	0	0	0	0	0	0	0	0	0	0
3.1.3	Working capital loan	147	0	0	0	0	147	0	0	0	0	0	0	0	0	0	0	0	0	0
3.1.4	Bond	0	0	0	0	0	0	0	0	0	0	0	0	0	0	0	0	0	0	0
3.1.5	Short term loan	0	0	0	0	0	0	0	0	0	0	0	0	0	0	0	0	0	0	0
3.1.6	Other inflow	0	0	0	0	0	0	0	0	0	0	0	0	0	0	0	0	0	0	0
3.2	Cash outflow	266470	531	1844	3507	5459	7280	12389	12389	12389	12389	12389	12389	12389	12389	12389	12389	12389	12389	12389
3.2.1	Various interest expenditure	120334	531	1844	3507	5459	7280	8264	8031	7785	7525	7250	6960	6653	6329	5987	5625	5243	4839	4412
3.2.2	Repayment of debt principal	146137	0	0	0	0	0	4125	4358	4604	4864	5139	5429	5736	6060	6402	6764	7146	7550	7976
3.3.3	Profit payable (dividend distribution)	0	0	0	0	0	0	0	0	0	0	0	0	0	0	0	0	0	0	0
3.3.4	Other outflows	0	0	0	0	0	0	0	0	0	0	0	0	0	0	0	0	0	0	0
4	Net cash flow	173481	0	0	0	0	0	3115	3115	2904	2871	2837	2486	2409	2328	2243	2152	2057	1956	1849
5	Accumulated surplus funds	165416	0	0	0	0	0	3115	6231	9135	12006	14843	17329	19739	22067	24310	26462	28519	30475	32324

9.4　Financial Evaluation

Continued

Serial number	Item	Operating period																
		19	20	21	22	23	24	25	26	27	28	29	30	31	32	33	34	35
1	Net cash flow from operating activities	14125	14006	13881	13748	13607	13459	13302	13137	13137	13137	13137	13137	13137	13137	13137	13137	13593
1.1	Cash inflow	19818	19818	19818	19818	19818	19818	19818	19818	19818	19818	19818	19818	19818	19818	19818	19818	19818
1.1.1	Sales income	19818	19818	19818	19818	19818	19818	19818	19818	19818	19818	19818	19818	19818	19818	19818	19818	19818
1.1.2	Substituted money on VAT	0	0	0	0	0	0	0	0	0	0	0	0	0	0	0	0	0
1.1.3	Subsidy income	0	0	0	0	0	0	0	0	0	0	0	0	0	0	0	0	0
1.1.4	Other inflows	0	0	0	0	0	0	0	0	0	0	0	0	0	0	0	0	0
1.2	Cash outflows	5693	5812	5938	6070	6211	6359	6516	6681	6681	6681	6681	6681	6681	6681	6681	6681	6225
1.2.1	Operating cost	3977	3977	3977	3977	3977	3977	3977	3977	3977	3977	3977	3977	3977	3977	3977	3977	3977
1.2.2	Input value added tax	0	0	0	0	0	0	0	0	0	0	0	0	0	0	0	0	0
1.2.3	Sales tax and surcharges	337	337	337	337	337	337	337	337	337	337	337	337	337	337	337	337	337
1.2.4	Value added tax	0	0	0	0	0	0	0	0	0	0	0	0	0	0	0	0	0
1.2.5	Income tax	1379	1498	1624	1756	1897	2045	2202	2367	2367	2367	2367	2367	2367	2367	2367	2367	1911
1.2.6	Other outflows	0	0	0	0	0	0	0	0	0	0	0	0	0	0	0	0	0
2	Net cash flow from investment activities	0	0	0	0	0	0	0	0	0	0	0	0	0	0	0	0	0
2.1	Cash inflow	0	0	0	0	0	0	0	0	0	0	0	0	0	0	0	0	0
2.2	Cash outflow	0	0	0	0	0	0	0	0	0	0	0	0	0	0	0	0	0
2.2.1	Construction investment	0	0	0	0	0	0	0	0	0	0	0	0	0	0	0	0	0
2.2.2	Maintain operation investment	0	0	0	0	0	0	0	0	0	0	0	0	0	0	0	0	0

9 Case Study of Economic Evaluation of Hydropower Projects

Continued

Serial number	Item	\multicolumn{17}{c}{Operating period}																
		19	20	21	22	23	24	25	26	27	28	29	30	31	32	33	34	35
2.2.3	Working capital	0	0	0	0	0	0	0	0	0	0	0	0	0	0	0	0	0
2.2.4	Other outflows	0	0	0	0	0	0	0	0	0	0	0	0	0	0	0	0	0
3	Net cash flow from financing activities	−12389	−12389	−12389	−12389	−12389	−12389	−12389	−7	−7	−7	−7	−7	−7	−7	−7	−7	−7
3.1	Cash inflow	0	0	0	0	0	0	0	0	0	0	0	0	0	0	0	0	0
3.1.1	Project capital inflow	0	0	0	0	0	0	0	0	0	0	0	0	0	0	0	0	0
3.1.2	Construction investment loan	0	0	0	0	0	0	0	0	0	0	0	0	0	0	0	0	0
3.1.3	Working capital loan	0	0	0	0	0	0	0	0	0	0	0	0	0	0	0	0	0
3.1.4	Bond	0	0	0	0	0	0	0	0	0	0	0	0	0	0	0	0	0
3.1.5	Short term loan	0	0	0	0	0	0	0	0	0	0	0	0	0	0	0	0	0
3.1.6	Other inflows	0	0	0	0	0	0	0	0	0	0	0	0	0	0	0	0	0
3.2	Cash outflow	12389	12389	12389	12389	12389	12389	12389	7	7	7	7	7	7	7	7	7	7
3.2.1	Various interest expenditure	3962	3486	2982	2451	1890	1296	670	7	7	7	7	7	7	7	7	7	7
3.2.2	Repayment of debt principal	8427	8903	9406	9938	10499	11092	11719	0	0	0	0	0	0	0	0	0	0
3.3.3	Profit payable (dividend distribution)	0	0	0	0	0	0	0	0	0	0	0	0	0	0	0	0	0
3.3.4	Other outflows	0	0	0	0	0	0	0	0	0	0	0	0	0	0	0	0	0
4	Net cash flow	1737	1618	1492	1359	1219	1070	914	13129	13129	13129	13129	13129	13129	13129	13129	13129	13586
5	Accumulated surplus funds	34061	35678	37170	38529	39747	40818	41731	54861	67990	81119	94248	107378	120507	133636	146765	159895	173481

9.4 Financial Evaluation

Table 9.10 Financial cash flow statement of project investment

unit: 10^4 yuan

Serial number	Item	Total	Construction period					Operation period											
			1	2	3	4	5	6	7	8	9	10	11	12	13	14	15	16	17	18
1	Cash inflow	594753	0	0	0	0	0	19818	19818	19818	19818	19818	19818	19818	19818	19818	19818	19818	19818	19818
1.1	Power generation sales income	594543	0	0	0	0	0	19818	19818	19818	19818	19818	19818	19818	19818	19818	19818	19818	19818	19818
1.2	Subsidy income and other income	0	0	0	0	0	0	0	0	0	0	0	0	0	0	0	0	0	0	0
1.3	Project residual value	0	0	0	0	0	0	0	0	0	0	0	0	0	0	0	0	0	0	0
1.4	Recovery of working capital	210	0	0	0	0	0													
2	Cash outflow	293653	23607	33773	36550	43286	27019	4314	4314	4314	4314	4314	4314	4314	4314	4314	4314	4314	4314	4314
2.1	Construction investment	164024	23607	33773	36550	43286	26809	0	0	0	0	0	0	0	0	0	0	0	0	0
2.2	Working capital	210	0	0	0	0	210	0	0	0	0	0	0	0	0	0	0	0	0	0
2.3	Operating cost	119312	0	0	0	0	0	3977	3977	3977	3977	3977	3977	3977	3977	3977	3977	3977	3977	3977
2.4	Sales tax and surcharges	10107	0	0	0	0	0	337	337	337	337	337	337	337	337	337	337	337	337	337
2.5	Maintain operation investment	0	0	0	0	0	0	0	0	0	0	0	0	0	0	0	0	0	0	0
3	Net cash flow before income tax	301099	−23607	−33773	−36550	−43286	−27019	15504	15504	15504	15504	15504	15504	15504	15504	15504	15504	15504	15504	15504
4	Accumulated net cash flow before income tax	1806237	−23607	−57380	−93929	−137215	−164234	−148730	−133226	−117722	−102218	−86713	−71209	−55705	−40201	−24697	−9193	6311	21815	37319
5	Adjustment of income tax	70620	0	0	0	0	0	2369	2369	2369	2369	2369	2369	2369	2369	2369	2369	2369	2369	2369
6	Net cash flow after income tax	230480	−23607	−33773	−36550	−43286	−27019	13135	13135	13135	13135	13135	13135	13135	13135	13135	13135	13135	13135	13135
7	Net cash flow after accumulated income tax		−23607	−57380	−93929	−137215	−164234	−151099	−137964	−124829	−111694	−98559	−85425	−72290	−59155	−46020	−32885	−19750	−6615	6520

Continued

Serial number	Item	Operation period																
		19	20	21	22	23	24	25	26	27	28	29	30	31	32	33	34	35
1	Cash inflow	19818	19818	19818	19818	19818	19818	19818	19818	19818	19818	19818	19818	19818	19818	19818	19818	20028
1.1	Power generation sales revenue	19818	19818	19818	19818	19818	19818	19818	19818	19818	19818	19818	19818	19818	19818	19818	19818	19818
1.2	Subsidy income and other income	0	0	0	0	0	0	0	0	0	0	0	0	0	0	0	0	0
1.3	Project residual value	0	0	0	0	0	0	0	0	0	0	0	0	0	0	0	0	0
1.4	Recovery of working capital	0	0	0	0	0	0	0	0	0	0	0	0	0	0	0	0	210
2	Cash outflow	4314	4314	4314	4314	4314	4314	4314	4314	4314	4314	4314	4314	4314	4314	4314	4314	4314
2.1	Construction investment	0	0	0	0	0	0	0	0	0	0	0	0	0	0	0	0	0
2.2	Working capital	0	0	0	0	0	0	0	0	0	0	0	0	0	0	0	0	0
2.3	Operating cost	3977	3977	3977	3977	3977	3977	3977	3977	3977	3977	3977	3977	3977	3977	3977	3977	3977
2.4	Sales tax and surcharges	337	337	337	337	337	337	337	337	337	337	337	337	337	337	337	337	337
2.5	Maintain operation investment	0	0	0	0	0	0	0	0	0	0	0	0	0	0	0	0	0
3	Net cash flow become tax	15504	15504	15504	15504	15504	15504	15504	15504	15504	15504	15504	15504	15504	15504	15504	15504	15714
4	Accumulated net cash flow before income tax	52824	68328	83832	99336	114840	130344	145848	161352	176856	192361	207865	223369	238873	254377	269881	285385	301099
5	Adjustment of income tax	2369	2369	2369	2369	2369	2369	2369	2369	2369	2369	2369	2369	2369	2369	2369	2369	1913
6	Net cash flow after income tax	13135	13135	13135	13135	13135	13135	13135	13135	13135	13135	13135	13135	13135	13135	13135	13135	13802
7	Net cash flow after accumulated income tax	19655	32790	45924	59059	72194	85329	98464	111599	124734	137869	151004	164139	177273	190408	203543	216678	230480

Evaluation indicators:

	After income tax	Before income tax
FIRR of project investment	5.87%	7.22%
FNPV of project investment ($i_c = 8\%$)	−29342	−11217
Payback period of project investment/a	17.5	15.6

9.4 Financial Evaluation

Table 9.11　Financial cash flow statement of project capital

unit: 10^4 yuan

Serial number	Item	Total	Construction period					Operation period												
			1	2	3	4	5	6	7	8	9	10	11	12	13	14	15	16	17	18
1	Cash inflow	594606	0	0	0	0	0	19818	19818	19818	19818	19818	19818	19818	19818	19818	19818	19818	19818	19818
1.1	Power generation sales revenue	594543	0	0	0	0	0	19818	19818	19818	19818	19818	19818	19818	19818	19818	19818	19818	19818	19818
1.2	Subsidy income and other income	0	0	0	0	0	0	0	0	0	0	0	0	0	0	0	0	0	0	0
1.3	Project residual value	0	0	0	0	0	0	0	0	0	0	0	0	0	0	0	0	0	0	0
1.4	Recovery of working capital	63	0	0	0	0	0	0	0	0	0	0	0	0	0	0	0	0	0	0
2	Cash outflow	457408	4827	7123	8011	9749	6860	16695	16695	16907	16939	16974	17324	17401	17482	17568	17658	17754	17855	17961
2.1	Project capital fund	36571	4827	7123	8011	9749	6860	0	0	0	0	0	0	0	0	0	0	0	0	0
2.2	Repayment of loan principal	146137	0	0	0	0	0	4125	4358	4604	4864	5139	5429	5736	6060	6402	6764	7146	7550	7976
2.3	Loan interest payment	101488	0	0	0	0	0	8257	8024	7777	7517	7243	6952	6645	6321	5979	5617	5235	4831	4405
2.4	Operating cost	119312	0	0	0	0	0	3977	3977	3977	3977	3977	3977	3977	3977	3977	3977	3977	3977	3977
2.5	Sales tax and surcharges	10107	0	0	0	0	0	337	337	337	337	337	337	337	337	337	337	337	337	337
2.6	Income tax	43793	0	0	0	0	0	0	0	211	244	278	629	706	787	873	963	1059	1159	1266
3	Net cash flow	137198	−4827	−7123	−8011	−9749	−6860	3123	3123	2911	2879	2845	2494	2417	2336	2250	2160	2064	1963	1857
4	Accumulated net cash flow	457568	−4827	−11951	−19962	−29711	−36571	−33448	−30325	−27414	−24535	−21690	−19197	−16780	−14444	−12194	−10034	−7969	−6006	−4149

9 Case Study of Economic Evaluation of Hydropower Projects

Continued

Serial number	Item	\multicolumn{17}{c}{Operation period}
		19	20	21	22	23	24	25	26	27	28	29	30	31	32	33	34	35
1	Cash inflow	19818	19818	19818	19818	19818	19818	19818	19818	19818	19818	19818	19818	19818	19818	19818	19818	19881
1.1	Power generation sales revenue	19818	19818	19818	19818	19818	19818	19818	19818	19818	19818	19818	19818	19818	19818	19818	19818	19818
1.2	Subsidy income and other income	0	0	0	0	0	0	0	0	0	0	0	0	0	0	0	0	0
1.3	Project residual value	0	0	0	0	0	0	0	0	0	0	0	0	0	0	0	0	0
1.4	Recovery of working capital	0	0	0	0	0	0	0	0	0	0	0	0	0	0	0	0	63
2	Cash outflow	18074	18193	18319	18452	18592	18740	18897	6681	6681	6681	6681	6681	6681	6681	6681	6681	6225
2.1	Project capital fund	0	0	0	0	0	0	0	0	0	0	0	0	0	0	0	0	0
2.2	Repayment of loan principal	8427	8903	9406	9938	10499	11092	11719	0	0	0	0	0	0	0	0	0	0
2.3	Loan interest payment	3954	3478	2975	2444	1882	1289	662	0	0	0	0	0	0	0	0	0	0
2.4	Operating cost	3977	3977	3977	3977	3977	3977	3977	3977	3977	3977	3977	3977	3977	3977	3977	3977	3977
2.5	Sales tax and surcharges	337	337	337	337	337	337	337	337	337	337	337	337	337	337	337	337	337
2.6	Income tax	1379	1498	1624	1756	1897	2045	2202	2367	2367	2367	2367	2367	2367	2367	2367	2367	1911
3	Net cash flow	1744	1625	1499	1366	1226	1078	921	13137	13137	13137	13137	13137	13137	13137	13137	13137	13656
4	Accumulated net cash flow	−2405	−780	719	2085	3311	4389	5310	18447	31584	44721	57857	70994	84131	97268	110405	123541	137198

Evaluation indicator: FIRR of project capital: 8.00%

9.4 Financial Evaluation

Table 9.12　Balance sheet

unit: 10^4 yuan

Serial number	Item	Construction period					Operation period												
		1	2	3	4	5	6	7	8	9	10	11	12	13	14	15	16	17	18
1	Capital	24137	59754	99811	148556	182854	179943	177031	173907	170751	167561	164020	160402	156703	152919	149044	145073	141002	136824
1.1	Total amount of working capital	0	0	0	0	210	3325	6441	9345	12216	15053	17539	19949	22277	24520	26672	28729	30685	32534
1.1.1	Working capital	0	0	0	0	210	210	210	210	210	210	210	210	210	210	210	210	210	210
1.1.2	Accumulated surplus capital	0	0	0	0	0	3115	6231	9135	12006	14843	17329	19739	22067	24310	26462	28519	30475	32324
1.2	Project under construction	24137	59754	99811	148556	182644	0	0	0	0	0	0	0	0	0	0	0	0	0
1.3	Net value of fixed asset					0	176617	170590	164563	158535	152508	146481	140454	134426	128399	122372	116344	110317	104290
1.4	Net value of intangible asset and other capital																		
2	Liabilities and owner's equity	24137	59754	99811	148556	182854	179943	177031	173907	170751	167561	164020	160402	156703	152919	149044	145073	141002	136824
2.1	Total amount of current liabilities	0	0	0	0	0	0	0	0	0	0	0	0	0	0	0	0	0	0
2.2	Construction investment loan	19310	47803	79849	118845	146137	142012	137654	133051	128187	123048	117619	111883	105823	99421	92657	85511	77961	69985
2.3	Working capital loan	0	0	0	0	147	147	147	147	147	147	147	147	147	147	147	147	147	147
2.4	Subtotal of loans	19310	47803	79849	118845	146284	142159	137801	133198	128334	123195	117766	112030	105970	99568	92804	85658	78108	70132
2.5	Owner's equity	4827	11951	19962	29711	36571	37784	39229	40710	42418	44366	46254	48372	50733	53350	56240	59415	62894	66692
2.5.1	Capital fund	4827	11951	19962	29711	36571	36571	36571	36571	36571	36571	36571	36571	36571	36571	36571	36571	36571	36571
2.5.2	Capital reserve																		
2.5.3	Accumulated surplus reserve fund	0	0	0	0	0	121	266	414	585	780	968	1180	1416	1678	1967	2284	2632	3012
2.5.4	Accumulated undistributed profits	0	0	0	0	0	1091	2392	3725	5262	7016	8715	10621	12746	15102	17702	20560	23690	27109
	Calculation indicators: asset liability ratio/%	80.0	80.0	80.0	80.0	80.0	79.0	77.8	76.6	75.2	73.5	71.8	69.8	67.6	65.1	62.3	59.0	55.4	51.3

9 Case Study of Economic Evaluation of Hydropower Projects

Continued

Serial number	Item	Operation period																
		19	20	21	22	23	24	25	26	27	28	29	30	31	32	33	34	35
1	Capital	132533	128124	123588	118920	114111	109154	104040	111142	118244	125346	132448	139550	146652	153754	160856	167958	173691
1.1	Total amount working capital	34271	35888	37380	38739	39957	41028	41941	55071	68200	81329	94458	107588	120717	133846	146975	160105	173691
1.1.1	Working capital	210	210	210	210	210	210	210	210	210	210	210	210	210	210	210	210	210
1.1.2	Accumulated surplus capital fund	34061	35678	37170	38529	39747	40818	41731	54861	67990	81119	94248	107378	120507	133636	146765	159895	173481
1.2	Project under construction	0	0	0	0	0	0	0	0	0	0	0	0	0	0	0	0	0
1.3	Net value of fixed assets	98263	92235	86208	80181	74154	68126	62099	56072	50045	44017	37990	31963	25936	19908	13881	7854	
1.4	Net value of intangible and other assets																	
2	Liabilities and owner's equity	132533	128124	123588	118920	114111	109154	104040	111142	118244	125346	132448	139550	146652	153754	160856	167958	173691
2.1	Total amount of current liabilities	0	0	0	0	0	0	0	0	0	0	0	0	0	0	0	0	0
2.2	Construction investment loan	61558	52655	43248	33311	22811	11719	0	0	0	0	0	0	0	0	0	0	0
2.3	Working capital loan	147	147	147	147	147	147	147	147	147	147	147	147	147	147	147	147	147
2.4	Subtotal of loans	61705	52802	43395	33458	22958	11866	147	147	147	147	147	147	147	147	147	147	147
2.5	Owner's equity	70828	75322	80193	85462	91153	97288	103893	110995	118097	125199	132301	139403	146505	153607	160709	167811	173544
2.5.1	Capital fund	36571	36571	36571	36571	36571	36571	36571	36571	36571	36571	36571	36571	36571	36571	36571	36571	36571
2.5.2	Capital reserve																	
2.5.3	Accumulated surplus reserves	3426	3875	4362	4889	5458	6072	6732	7442	8153	8863	9573	10283	10993	11704	12414	13124	13697
2.5.4	Accumulated undistributed profits	30832	34876	39260	44002	49124	54645	60590	66982	73374	79766	86157	92549	98941	105333	111725	118117	123275
	Calculation indicators: asset liability ratio/%	46.6	41.2	35.1	28.1	20.1	10.9	0.1	0.1	0.1	0.1	0.1	0.1	0.1	0.1	0.1	0.1	0.1

9.5 Conclusions and Suggestions

9.4.5 Sensitivity Analysis

The main sensitive factors are construction investment, valid energy and loan interest rate. Herein, single-factor sensitivity analysis method is adopted here, and the variation ranges of the three factors are all set at ±10%. In addition, the sensitive impacts of loan repayment period and financial internal rate of return (FIRR) of project investment are also considered. And the results of sensitivity analysis are provided in Table 9.13.

Table 9.13 Sensitivity analysis of financial evaluation

Serial number	Item	FIRR/%		On-grid price during operation period/ [yuan/(kW·h)]	Payback period of loan/a	Payback time, PBT/a
		Project investment	Capital fund			
1	Basic plan	5.87	8	0.296	25	17.5
2	Variation of construction investment					
2.1	Increase by 10%	5.87	8	0.323	25	17.5
2.2	Decrease by 10%	5.87	8	0.269	25	17.5
3	Variation of valid energy					
3.1	Increase by 10%	5.87	8	0.270	25	17.5
3.2	Decrease by 10%	5.87	8	0.328	25	17.5
4	Variation of loan rate					
4.1	Decrease by 10%	6.18	8	0.307	25	17
4.2	Decrease by 10%	5.56	8	0.286	25	18
5	Variation of loan payment period					
5.1	Twenty years	6.2	8	0.307	20	17
6	FIRR of project investment, 8%	8	15.1	0.374	25	14.7

Note: The electricity price in the above table, if not specified, is the electricity price excluding VAT.

For Q Hydropower Station, as can be seen from the above table, the financial indicators suffer little from the sensitive factors, with small fluctuations. During the operation period, the on-grid price varies between 0.269–0.374 yuan/(kW·h). Compared with the basic scheme, the maximum adverse variation is about 26.4%. It indicates that the project has some anti-risk ability.

9.5 Conclusions and Suggestions

For Q Hydropower Station: EIRR is 9.7%, higher than the social discount rate of

8%. ENPV is 175.59 million yuan, larger than zero. The economic benefit cost ratio (EBCR) is 1.11, larger than 1.0. These indicate that the project is economic and feasible.

The capital of this project is 365.71 million yuan, and the estimated on-grid price is 0.296 yuan/(kW·h) when FIRR is 8%. The FIRR of project investment after income tax is 5.87%. The total profit from power generation is 1807.66 million yuan. The investment payoff period is 17.5 years. Above all, the project is financially feasible.

10

Case Study of Economic Evaluation of Water Conservancy Projects

10.1 Project Overview

H Reservoir is in ××.Province. The project provides many benefits in addition to water supply: flood control, power generation, and other comprehensive utilization. The project is mainly composed of water retaining structure, release structure, water diversion system and powerhouse. The normal storage water level and dead storage level of the reservoir is 210 m and 180 m, respectively. The restricted water level in flood season is 205 m. The upper water level for flood control is 211 m. The check flood level is 213.1 m. The total storage capacity and regulating storage capacity of the reservoir is 290 million m³ and 210 million m³, respectively. The flood control storage capacity is 80 million m³, of which the combined storage capacity of flood control and profit promotion is 60 million m³, and the special storage capacity for flood control is 20 million m³. The reservoir can improve the flood control standard of the downstream river from once in 10 years to once in 30 years. The average annual water supply is 156 million m³. The installed capacity of the power station is 60 MW. The average annual power generation is 160 million kW·h. The effective power utilization coefficient is 1. The power consumption rate is 0.2%, and the power transmission and transformation loss rate is 2%. The on-grid power is 156.84 million kW·h. The total construction period of this project is 4 years. See Table 10.1 for the unit operation plan.

Table 10.1　　　　　　　　　　Unit commissioning schedule

Serial number	Item	1	2	3	4	Fifth year and beyond
1	Installed capacity at the end of year/MW	0	0	0	60	60
2	Annual power generation/(10^8 kW·h)					1.60
3	Effective energy/(10^8 kW·h)					1.60

Continued

Serial number	Item	1	2	3	4	Fifth year and beyond
4	Plant energy/(10^8 kW·h)					1.5968
5	On-grid energy/(10^8 kW·h)					1.5684

10.2　Evaluation Purpose and Basis

10.2.1　Purpose of Economic Evaluation

In accordance with the relevant laws, regulations and current financial system of the government, the economic evaluation of the project aims at the two following aspects: Evaluate the economic feasibility of the project, including the analysis of the economic benefits and costs of the project, and calculate the economic indicators of the project. On the other hand, evaluate the financial feasibility of the project, including the calculation of the project cost, the analysis of the project loan capacity as well as the proposal of the project financing plan, etc.

10.2.2　Evaluation Basis

(1) *Water industry policy* (GF [1997] No. 35).

(2) *Notice of the State Council on trial implementation of capital system for fixed asset investment projects* (GF [1996] No. 35).

(3) *Construction Projects Economic Evaluation Method and Parameter* (3rd Edition) (*fgtz* [2006] No. 1325).

(4) *Regulation for Economic Evaluation of Water Conservancy Construction Projects* (SL 72-2013).

(5) *Measures for Price Administration of Water Supply of Water Engineering* (*No. 4* [2003], NDRC and MWR).

10.3　Fixed Assets Investment and Total Cost

10.3.1　Fixed Assets Investment

The fixed-asset investment of the project is 2033.57 million yuan. Specifically, the investment of the key project is 1192.06 million yuan. And the compensation fee for the reservoir inundation treatment together with the investment in environmental impact assessment and water and soil conservation is 841.51 million yuan. The annual investment of the project can be seen in Table 10.2.

10.3 Fixed Assets Investment and Total Cost

Table 10.2　　　　　　　　Annual investment of project items　　　　　unit: 10^4 yuan

Project Items	Annual Year				
	1	2	3	4	Total
Key Project	27595	36487	31985	23139	119206
Project Immigrant & Environment Protection	21141	20725	20935	21350	84151
Total	48736	57212	52920	44489	203357

10.3.2 Total Cost

To estimate the total cost of the project, the factor of production cost method is adopted. It is based on the item of the factor of production. The amortization of intangible assets and other assets are not considered. In details, the total cost includes repair cost, employee compensation, material cost, management cost, insurance cost, water resource cost, reservoir fund, depreciation cost, financial cost, etc. They are calculated as follows:

1. Repair cost

Repair cost is required for the normal operation and use of the fixed assets, including maintenance of high efficiency of use and fixed repairs. According to the size of the repair scope and the length of the repair time interval, it can be divided into major, minor and medium repair costs. Considering the characteristics of the project, the annual maintenance cost is estimated as 1% of the value of fixed assets (deducting compensation for land acquisition by immigrants), and the annual operation cost before financing is 11.92 million yuan.

2. Employee compensation

Employee compensation is various forms of remuneration and other related expenses in order to obtain the services provided by employees. It includes employee wages (including wages, bonuses, allowances, subsidies and other monetary rewards), employee welfare expenses, trade union funds, employee education funds, housing fund, and some basic social insurance premiums, including medical insurance premium, endowment insurance premium, unemployment insurance premium, work injury insurance, maternity insurance, etc.

Assume the personnel quota of the project is 80. The average annual salary is determined as 50,000 yuan, given the investigation of that in recent three years. Therefore, the total annual salary is 4 million yuan. The welfare expenses are estimated as 62% of the salary, including welfare expenses (14%), the medical insurance premiums (9%), the endowment insurance premiums (20%), the unemployment insurance premiums (2%), the work injury insurance premiums (1.5%), the maternity insurance premiums (1%), the housing fund (10%), the union funds (2%), and education funds (2.5%). Therefore, the annual salary is 6.48 million yuan.

3. Material expenses

Material expenses include the costs of raw materials, raw water, auxiliary materials, spare parts that consumed in the operation and maintenance of water conservancy projects. Assuming as 0.1% of the original value of fixed assets (excluding compensation for land acquisition by immigrants), the material cost of the year before financing is 1.19 million yuan.

4. Other expenses

Other expenses are related to water supply production and operation in the operation and maintenance of water conservancy projects, except for salaries, material expenses, etc. Specifically, engineering observation expenses, water quality testing expenses, temporary facility expenses are included. Assuming as 10% of the total of 1, 2 and 3, other expenses in the year before financing are 1.96 million yuan.

5. Management fee

The management fee includes the expenses for travel, office, consulting, audit, litigation, pollution discharge, greening, business entertainment, bad debt losses of the management organization of the water conservancy project. Herein, assuming as one time of the annual salary, the estimated annual management fee is 6.48 million yuan.

6. Premiums

The insurance premium includes fixed assets insurance and other insurance. Assuming as 0.25% of the original value of fixed assets (excluding compensation for land acquisition for immigrants), the annual insurance premium before financing is 2.98 million yuan.

7. Water resources fee

According to the local documents, the water resources fee for hydropower generation is 0.008 yuan/kW·h. The water resources fee for urban and rural living and industrial water supply is 0.07 yuan/m^3. Given that the annual power generation is 160 million kW·h, the annual water resources fee for power generation is 1.28 million yuan. The annual water supply is 156 million m^3. In addition, the annual water resources fee for water supply is 10.92 million yuan. To sum up, the total annual water resources fee is 12.2 million yuan.

8. Reservoir area fund

The reservoir area fund is the required expenses for these: support the implementation of the infrastructure construction and economic development planning of the reservoir area and the resettlement area; support the reservoir area protection project and the production and maintenance of living facilities; and solve other problems of the reservoir immigrants after the reservoir impoundment.

According to *Interim Measures for the Use and Management of the Fund in the Reservoir Area of a Large and Medium* formulated by the Ministry of Finance, the state integrates the fund supporting the former reservoir area maintenance, post-support fund for the original reservoir area and the post-support fund for the resettlement of the operational large and medium-sized reservoirs (installed capacity of 25000 kW and above, with generation income), and establishes the funds for the large and medium-sized reservoir area (i.e., the "reservoir area fund"). The reservoir area fund is raised from the power generation income of large and medium-sized reservoirs, according to the actual online sales of electricity at the standard price of 0.008 yuan/kW · h. The annual power generation of this project is 156 million kW · h. And the annual reservoir area fund is calculated as 1.25 million yuan.

9. Depreciation expenses

Depreciation expenses refer to the depreciation expense of fixed assets. The fixed assets of the project will be worn out in the use process, and the loss of value is usually compensated by depreciation. The straight-line method for depreciation calculation is adopted, and the residual value is 4%. According to the depreciation life of the project, the depreciation life of the reservoir dam and the electromechanical equipment is 50 years and 25 years, respectively. Assuming the comprehensive depreciation life of the project as 40 years, and the annual average comprehensive depreciation rate is 2.4%.

Depreciation expenses = (*value of fixed assets* − *residual value of the project*)
×*comprehensive depreciation rate*

The annual depreciation expenses of the project are 48.81 million yuan.

10. Financial expenses

Financial expenses are composed of long-term loan interest, short-term loan interest and current loan interest. Long-term loan interest is calculated based on compound interest. Interest during construction period is included in the value of fixed assets of the project, while interest during operation period is included in production cost.

11. Annual operation cost

The annual operation cost is the total cost excluding depreciation, amortization and interest expenses. The annual operation cost includes repair expenses, salary, material expenses, other expenses, management fee, insurance premiums, water resources fee and reservoir area fund. The annual operation cost of the project before financing is 44.46 million yuan.

12. Total cost

According to the above estimation, the total cost of the project before financing is 93.27 million yuan.

10.3.3 Circulating Fund

Circulating fund is the operating fund which will be tied up for a long time for turnover use during the operation period. According to the estimation of 10% of the annual operation cost, the circulating fund of the project is 4.45 million yuan.

10.4 National Economic Evaluation

10.4.1 Engineering Cost

1. Fixed assets investment

The fixed asset investment of the project is 2033.57 million yuan. Specifically, the investment of the hub project is 1192.06 million yuan, and the investment of resettlement and environmental protection is 841.51 million yuan. The investment estimate of this project adopts the market price, therefore price adjustment is not needed in the national economic evaluation. But the taxes for the internal transfer payment should be eliminated. For this project, the total value of the taxes is 201.69 million yuan. Specifically, the construction tax is 36.11 million yuan, and the remaining 165.5 million yuan is composed of the cultivated land occupation tax, cultivated land reclamation fee and forest vegetation restoration fee. Therefore, the investment in fixed assets, deducting the internal transfer taxes of the national economy, is 1831.88 million yuan, as shown in Table 10.3.

Table 10.3　　　Annual investment of fixed assets after adjustment　　　unit: 10^4 yuan

Item	Year				
	1	2	3	4	Total
Key project	26759	35382	31016	22438	115595
Immigration environment protection	16981	16646	16816	17148	67593
Total	43740	52028	47832	39586	183188

2. Annual operation cost

The annual operation cost of the national economy is based on that of the financial evaluation, excluding the internal transfer, i.e., water resources fee. For this project, the estimated annual operation cost is 32.26 million yuan.

3. Working capital

The working capital is estimated as 10% of the annual operation cost. For this project, the estimate is 3.23 million yuan.

4. Renovation cost

For the mechanical and electrical equipment and metal structure equipment, the economic service life is 25 years. The equipment should be renewed in the economic calculation period. Estimate as 75% of the investment of metal mechanical and electrical equipment, the renewal and transformation cost of the project is 119.88 million yuan. Invested in two years, and the annual investment is 59.94 million yuan.

10.4.2 Engineering Benefit

The project mainly benefits from water supply, flood control and power generation.

1. Water supply benefits

This project will increase the water supply to the receiving area by 156 million m^3, providing a strong guarantee for the local economic and social development. The water supply benefit is calculated by sharing coefficient method. The water quota of industrial added value of ten thousand yuan in water receiving area is $26m^3/10^4$ yuan, and the added value generated by one cubic meter of water is 384.6 $yuan/m^3$.

The allocation coefficient is a comprehensive function which reflects the interrelations of the multiple factors, which are related to industry and water supply as well as affecting the benefits of industrial water supply. The allocation coefficient is influenced by many factors, including the level of industrial development, the status of water supply project, industrial structure, investment structure, price system, water saving level, management level, etc. The allocation coefficient can be calculated by investment ratio, fixed assets ratio, cost ratio, capital occupation ratio, and discount cost ratio, etc. Herein, the investment ratio method is adopted. Through investigation, the investment of industrial water supply projects (including water source, water transmission, water plants, pipe networks and other facilities) accounts for 5% of the investment in industrial production (including investment in industrial water supply). Therefore, the allocation coefficient is set as 5%. The calculated benefit of the industrial water supply project in the receiving area is 3000 million yuan.

After the completion of H Reservoir, water transmission project, water plant, water supply pipe network and other water supply facilities need to be built. The benefit sharing coefficient of this project for water supply and supporting projects is 0.1. The annual average water supply benefit is 300 million yuan. At the initial stage of the project operation, 30% of the designed water supply benefit is considered, i.e., 90 million yuan. The designed water supply benefit is achieved according to the 10-year growth.

2. Benefits of flood control

According to *Regulation for Economic Evaluation of Water Conservancy Construction Projects (SL 72 -2013)*, the multi-year average flood control benefits of flood control projects can be calculated according to the flood losses that can be reduced or not compared with the project, including direct flood control benefits and indirect flood control benefits. The flood control benefit of the project is calculated by the frequency method. The flood loss that can be reduced within the flood control scope after the project is completed can be regarded as the flood control benefit of the project.

After the completion of the project, the flood control standard of the downstream cities can be raised from the current 10-year return period to the 30-year return period.

10 Case Study of Economic Evaluation of Water Conservancy Projects

The current protected population is 150000, and the protected urban area is 13.9 km². In the design level year 2020, the protected population will be 280000, and the protected urban built-up area will be 25.9 km². In 2020, the industrial added value will be 17700 million yuan.

The direct flood control benefit is the deductible flood loss, which includes the loss caused by the inundation of population, the damage of houses and facilities, as well as the loss caused by the shutdown of industry and mining, the suspension of business and the interruption of traffic, power and communication. The reduced and reduced flood losses are calculated according to the flood reduction physical indicators, the multi-year average flood losses are calculated by the frequency method, and the losses caused by different frequency flood disasters are determined by the comprehensive analysis of the current loss and loss growth rate index according to the scope of flood inundation, the typical investigation method, and the calculation results of different frequency flood losses are shown in Table 10.4. It can be seen from Table 10.4 that the current flood control standard is once in 10 years, and the annual average flood control loss without project is 79.6 million yuan. After the completion of the project, the flood control standard is increased from once in 10 years to once in 30 years, and the annual average flood loss is 38.45 million yuan. The multi-year average flood control benefit of the project is calculated as 41.15 million yuan based on whether the project can be reduced or not. With the development of social economy,

Table 10.4   Annual average flood control benefit estimation

Project conditions	Flood frequency $P/\%$	Economic losses/10^8 yuan Sn	Frequency difference ΔP	Average economic loss/10^8 yuan $S=(S_1+S_2)/2$	Economic loss of adjacent frequency/10^4 yuan $\Delta P \times S$	Average annual loss/10^4 yuan $\Sigma \Delta P \times S$
Without project	>10	0				
	10	3.48				
	5	5.55	0.05	4.515	2258	2258
	3.33	9.8	0.0167	7.675	1282	3539
	2	11.5	0.0133	10.65	1416	4956
	1	15.6	0.01	13.55	1355	6311
	0.1	17.8	0.009	16.7	1503	7814
	0.02	18.9	0.0008	18.35	147	7960
With project	3.33	0				
	2	10	0.0133	8	1064	1064
	1	14.6	0.01	12.3	1230	2294
	0.1	16.8	0.009	15.7	1413	3707
	0.02	17.9	0.0008	17.35	138	3845

Note: Under project control, the flood loss is 600-million-yuan meeting with floods with frequency of 30 years or more.

10.4 National Economic Evaluation

the property of flood control protection object is gradually increasing, and the comprehensive growth rate of property is 3%.

The indirect economic loss that can be reduced by flood control project is called indirect flood control benefit. It mainly includes the loss caused by the interruption and shortage of raw materials and fuel supply of the industrial and mining enterprises in the flooded area, resulting in the shutdown, production reduction and cost increase of the enterprises; the economic loss caused by the failure of the production in the flooded area to return to normal soon after the flood. According to the typical investigation, the indirect flood control benefit of the project area is about 20% of the direct flood control benefit, so the indirect flood control benefit of the project is 8.23 million yuan.

The flood control benefit of the project is the sum of direct flood control benefit and indirect flood control benefit, which is 49.38 million yuan.

3. Power generation benefits

The project has a hydropower station with an installed capacity of 60 MW and an average annual generating capacity of 160 million kW·h. The cost calculation of alternative thermal power plant, which is the best equivalent alternative power plant for power generation benefit.

According to the results of electric power balance, the working capacity of the power station in the power grid is 60 MW, and the annual generating capacity is 160 million kW·h. The capacity coefficient of the alternative thermal power station is 1.1, the electricity coefficient is 1.05, the installed capacity of the alternative thermal power station is 66 MW, and the annual generating capacity of the alternative thermal power station is 176 million kW·h.

According to the data of new thermal power station, the construction investment per kilowatt is 4500 yuan/kW, the fixed operation rate of thermal power station is 4%, the coal consumption per kilowatt hour of thermal power station in the project area is 350 g/kW·h, and the standard coal price of power grid where the project is located is 400 yuan/t. The main parameters of alternative thermal power plants are as follows:

Investment per kilowatt: 4500 yuan/kW;

Coal consumption per kilowatt hour: 350g/(kW·h);

Standard coal price: 400 yuan/t;

Fixed operating rate: 4%.

According to the above parameters, it is calculated that the investment of alternative thermal power construction is 297 million yuan, the fuel cost is 35.28 million yuan, and the operating cost is 11.88 million yuan.

The investment of thermal power station project is divided into three years, with the investment proportion of 30%, 40% and 30%, respectively. After 25 years of operation, the thermal power station will be renovated. The cost of renovation is calculated as 75%

of the project investment, which is 222.76 million yuan, divided into two years.

10.4.3 Evaluation Methods of National Economy and the Results

According to *Construction Projects Economic Evaluation Method and Parameter (3rd Edition)*, the social discount rate of water conservancy projects is 8%.

The total construction and production periods are 4 years and 50 years, respectively. Therefore, the economic calculation period is 54 years. Take the first year of construction as the base year and the beginning of the year as the conversion starting point.

According to *Construction Projects Economic Evaluation Method and Parameter (3rd Edition)*, economic net present value, economic internal rate of return, benefit cost ratio, etc. are used as national economic evaluation indexes.

The economic service life of the mechanical and electrical equipment of the hydropower station is 25 years. When it is due, the equipment shall be renewed according to the original scale. The renewal cost shall be divided into two years, and the annual investment proportion shall be 50% and 50%, respectively.

The benefit cost flow of national economic evaluation is shown in Table 10.5.

The net benefit flow is calculated by IRR (value) of internal function in Excel (value is flow value, the same below) and the economic internal rate of return is 14.24%. The net benefit flow in Table 10.5 is calculated by NPV (0.08, value) of internal function in Excel and the economic net present value is 1577.89 million yuan. The alternative cost (benefit) flow in Table 10.5 is calculated by internal function in Excel NPV (0.08, value) is calculated to get benefit present value of 3405.37 million yuan. The design scheme cost flow in Table 10.5 is calculated to get cost present value of 1827.48 million yuan by using internal function NPV (0.08, value) in Excel, and the economic benefit cost ratio (benefit present value/cost present value) is calculated to be 1.86.

According to the above calculation results, the internal rate of return of the project economy is 14.24%, more than 8%, the economic net present value (ENPV) is 1577.89 million yuan, greater than zero, and the ratio of economic benefit to cost is 1.86, greater than 1. Above all, the project is economically reasonable.

10.4.4 Sensitivity Analysis

Sensitivity analysis is a kind of uncertainty analysis method to find out the sensitive factors which have important influence on the economic benefit index of investment project from many uncertain factors, and to analyze and calculate the influence degree and sensitivity degree on the economic benefit index of the project, and then to judge the risk tolerance of the project. The sensitivity factors of the project mainly include fixed asset investment, water supply, flood control benefit and effective electricity. The following eight situations are proposed for sensitivity analysis according to the possible floating range of single

10.4 National Economic Evaluation

Table 10.5 Statement of cost benefit flow

unit: 10^4 yuan

Serial number	Item	Calculation period																		
		1	2	3	4	5	6	7	8	9	10	11	12	13	14	15	16	17	18	19
1	Benefit flow	0	8910	11880	8910	18654	20902	23155	25412	27674	29940	32212	34489	36771	39059	41352	41551	41756	41968	42185
1.1	Project benefit	0	8910	11880	8910	18654	20902	23155	25412	27674	29940	32212	34489	36771	39059	41352	41551	41756	41968	42185
1.1.1	Water supply benefit	0	0	0	0	9000	11100	13200	15300	17400	19500	21600	23700	25800	27900	30000	30000	30000	30000	30000
1.1.2	Flood control benefit	0	0	0	0	4938	5086	5239	5396	5558	5724	5896	6073	6255	6443	6636	6835	7040	7252	7469
1.1.3	Power generation benefit	0	8910	11880	8910	4716	4716	4716	4716	4716	4716	4716	4716	4716	4716	4716	4716	4716	4716	4716
1.2	Recovery of residual value of fixed assets																			
1.3	Recovery of working capital																			
2	Cost flow	43740	52028	47832	40032	3226	3226	3226	3226	3226	3226	3226	3226	3226	3226	3226	3226	3226	3226	3226
2.1	Fixed-asset investment	43740	52028	47832	39587															
2.2	Annual operation cost					3226	3226	3226	3226	3226	3226	3226	3226	3226	3226	3226	3226	3226	3226	3226
2.3	Working capital				445															
2.4	Renovation cost																			
3	Net benefit flow	−43740	−43118	−35952	−31122	15428	17676	19929	22186	24448	26714	28986	31263	33545	35833	38126	38325	38530	38742	38959
4	Accumulated net benefit flow	−43740	−86858	−122810	−153932	−138504	−120828	−100899	−78713	−54266	−27551	1435	32698	66243	102076	140203	178528	217058	255800	294759

10　Case Study of Economic Evaluation of Water Conservancy Projects

Continued

Serial number	Item	Calculation period																		
		20	21	22	23	24	25	26	27	28	29	30	31	32	33	34	35	36	37	38
1	Benefit flow	42409	42640	42878	43123	43375	43635	43902	44178	44462	55892	56193	45365	45685	46014	46353	46702	47061	47432	47813
1.1	Project benefit	42409	42640	42878	43123	43375	43635	43902	44178	44462	55892	56193	45365	45685	46014	46353	46702	47061	47432	47813
1.1.1	Benefit of industrial water supply	30000	30000	30000	30000	30000	30000	30000	30000	30000	30000	30000	30000	30000	30000	30000	30000	30000	30000	30000
1.1.2	Flood control benefit	7693	7924	8162	8407	8659	8919	9186	9462	9746	10038	10339	10649	10969	11298	11637	11986	12345	12716	13097
1.1.3	Power generation benefit	4716	4716	4716	4716	4716	4716	4716	4716	4716	15854	15854	4716	4716	4716	4716	4716	4716	4716	4716
1.2	Recovery of residual value of fixed assets																			
1.3	Recovery of working capital																			
2	Cost flow	3226	3226	3226	3226	3226	3226	3226	3226	3226	9220	9220	3226	3226	3226	3226	3226	3226	3226	3226
2.1	Investment in fixed assets																			
2.2	Annual operation cost	3226	3226	3226	3226	3226	3226	3226	3226	3226	3226	3226	3226	3226	3226	3226	3226	3226	3226	3226
2.3	Working capital										5994									
2.4	Renovation fee											5994								
3	Net benefit flow	39183	39414	39652	39897	40149	40409	40676	40952	41236	46672	46973	42139	42459	42788	43127	43476	43835	44206	44587
4	Accumulated net benefit flow	333942	373357	413008	452905	493054	533462	574138	615090	656326	702998	749971	792110	834569	877356	920483	963959	1007794	1052000	1096587

216

10.4 National Economic Evaluation

Continued

Serial number	Item	Calculation period																Total
		39	40	41	42	43	44	45	46	47	48	49	50	51	52	53	54	
1	Benefit flow	48206	48611	49028	49457	49899	50355	50824	51307	51805	52318	52846	53390	53950	54527	55121	63506	2237040
1.1	Project benefit	48206	48611	49028	49457	49899	50355	50824	51307	51805	52318	52846	53390	53950	54527	55121	55733	2229267
1.1.1	Industrial water supply benefit	30000	30000	30000	30000	30000	30000	30000	30000	30000	30000	30000	30000	30000	30000	30000	30000	1384500
1.1.2	Flood control benefit	13490	13895	14312	14741	15183	15639	16108	16591	17089	17602	18130	18674	19234	19811	20405	21017	556991
1.1.3	Power generation benefit	4716	4716	4716	4716	4716	4716	4716	4716	4716	4716	4716	4716	4716	4716	4716	4716	287776
1.2	Recovery of residual value of fixed assets																7328	7328
1.3	Recovery of working capital																445	445
2	Expense flow	3226	3226	3226	3226	3226	3226	3226	3226	3226	3226	3226	3226	3226	3226	3226	3226	356920
2.1	Fixed asset investment																	183187
2.2	Annual operation cost	3226	3226	3226	3226	3226	3226	3226	3226	3226	3226	3226	3226	3226	3226	3226	3226	161300
2.3	Working capital																	445
2.4	Renovation cost																	11988
3	Net benefit flow	44980	45385	45802	46231	46673	47129	47598	48081	48579	49092	49620	50164	50724	51301	51895	60280	1880120
4	Accumulated net benefit flow	1141568	1186952	1232754	1278985	1325658	1372787	1420385	1468466	1517045	1566137	1615756	1665920	1716644	1767944	1819840	1880120	

Evaluation indicators: FIRR 14.24%

ENPV ($i_s = 8\%$) 1577.89 million yuan

EBCR ($i_s = 8\%$) 1.86

factor of fixed asset investment, water supply, flood control benefit and effective electricity:

(1) Fixed asset investment increased by 10%, and other factors remained unchanged.

(2) Fixed asset investment decreased by 10%, other factors remained unchanged.

(3) Water supply increased by 10%, and other factors remained unchanged.

(4) Water supply decreased by 10%, other factors remained unchanged.

(5) Flood control benefits increased by 10%, and other factors remained unchanged.

(6) The flood control benefit decreased by 10%, and other factors remained unchanged.

(7) The effective power increased by 10%, and other factors remained unchanged.

(8) The effective power is reduced by 10%, and other factors remain unchanged.

See Table 10.6 for the impact analysis results of various sensitive factors of the project on national economic evaluation indexes.

Table 10.6 Results of sensitivity analysis

Floating range	Indicator of national economy evaluation		
	FIRR/%	ENPV/10^4 yuan	EBCR
Basic plan	14.24	157789	1.86
Fixed assets increased 10%, others unchanged	13.10	139526	1.69
Fixed assets decreased 10%, others unchanged	15.60	176052	2.07
Water supply amount increased 10%, others unchanged	14.91	178417	1.98
Water supply decreased 10%, others unchanged	13.54	137161	1.75
Flood control benefit increased 10%, others unchanged	14.45	164370	1.90
Flood control benefit decreased 10%, others unchanged	14.02	151208	1.83
Effective energy increased 10%, others unchanged	14.58	164622	1.90
Effective energy decreased 10%, others unchanged	13.91	150956	1.83

The sensitivity analysis results show that when the fixed asset investment increases by 10% or the water supply, flood control benefit and effective electricity quantity decreases by 10%, the economic internal rate of return is 8% higher than the benchmark rate of return, and the economic net present value is greater than zero, which indicates that the project is economically reasonable and has strong anti-risk ability.

10.5 Cost Sharing and Cost Estimation

10.5.1 Investment in Shared Projects and Special Projects

The fixed asset investment of H reservoir is 2033.57 million yuan. According to the characteristics of the project, the fixed asset investment can be divided into two parts: special investment for power generation and common investment. The special investment for power generation includes the investment for power generation and diversion system, power generation plant, field area and switch station construction engineering, mechanical and electrical equipment and installation engineering. The special investment for power generation is 226 million yuan. Shared investment refers to fixed asset investment after deducting special investment, which is 1807.57 million yuan.

10.5.2 Principle of Cost Sharing

The task of this project is to take water supply as the main task, combine flood control with power generation and other comprehensive utilization. The principle of investment sharing is that the project investment specially serving a certain function shall be borne by this department alone, and the investment of shared projects jointly served by several departments shall be shared among these departments according to a certain ratio.

10.5.3 Investment Apportionment

Due to the complexity of the investment allocation of the comprehensive utilization hub, it is difficult to find a generally recognized allocation calculation method applicable to all kinds of comprehensive utilization water conservancy construction projects at present. Combined with the characteristics of the project, the proportion of storage capacity, the primary and secondary position method and the proportion of benefit present value method are to be considered for the investment allocation of fixed assets.

1. Proportion of storage capacity required by each function and allocation of primary and secondary positions

H reservoir has a storage capacity of 280 million m^3 below the normal pool level, a regulating storage capacity of 210 million m^3, a dead storage capacity of 50 million m^3, a flood control storage capacity of 80 million m^3, including a combined storage capacity of 60 million m^3 for flood control and profit promotion, and a dedicated storage capacity for flood control of 20 million m^3.

The task of H Reservoir is water supply, flood control and power generation. The storage capacity between the dead water level and the flood limited water level is 150 million m^3, which is set to meet the requirements of water supply and power generation. The storage capacity between the normal water level and the flood high water level is 20 million m^3, which is set to meet the requirements of flood control. The dead storage capacity is 50 million m^3, which is mainly set for the use of water supply and power generation. The storage capacity between the flood limited water level and the normal water level is 60 million m^3, which is set to meet the requirements of water supply and power generation. The storage capacity is shared by flood control, water supply and power generation.

According to the above functions, the idea of reservoir capacity allocation is to consider the water supply and power generation functions, allocate them with flood control functions, and then allocate the storage capacity occupied by water supply and power generation functions.

Among the above storage capacities, the dead storage capacity and the storage capacity between the dead water level and the flood limited water level of the reservoir are used for water supply and power generation, which are undertaken by water supply and power generation. The storage capacity between the normal water level and the flood high water level is dedicated for flood control, which is undertaken by flood control. The storage capacity between the flood limited water level and the normal water level is undertaken by the flood control, water supply and power generation departments, which operate according to the reservoir regulation Method: the storage capacity is used for flood control for 5 months and for water supply and power generation for 7 months. Although the use time of flood control storage capacity is short, it has a great impact on water supply and power generation. Flood control is in the leading position, sharing 80% of the storage capacity, which is 48 million m^3. Water supply and power generation are in the secondary position, sharing 20% of the storage capacity, which is 12 million m^3. Thus, the flood control shared storage capacity is $20+48=68$ million m^3. And the water supply and power generation shared storage capacity are $150+50+12=212$ million m^3.

Among the allocated storage capacity of water supply and power generation of 212 million m^3, considering that the power generation can use this storage capacity to raise the head to generate power when the water supply is satisfied, it is in a subordinate position. 20% of the allocated storage capacity of power generation is 42.4 million m^3, and the water supply is in a leading position, 80% of the allocated storage capacity is 169.6 million m^3.

According to the above sharing results, the shared storage capacity of water supply, flood control and power generation is 169.6 million m^3, 68 million m^3 and 42.4 million m^3, respectively. The corresponding proportions are 60.6%, 24.3% and 15.1%, respectively. According to the storage capacity ratio, the allocation investment of water supply, flood control, and power generation is 1095.39 million yuan, 439.24 million yuan and 272.94 million yuan, respectively. Considering the investment of special power generation projects, they are 1095.39 million yuan, 439.24 million yuan and 498.94 million yuan, respectively. The corresponding percentages are 53.9%, 21.6% and 24.5%, respectively of the fixed assets investment as shown in Table 10.7.

2. The proportion of benefit present value

When the proportion of benefit present value is used for fixed asset investment allocation, since the project benefit calculated in the national economic evaluation is generated by the common project and the special project together, this time the benefit present value is used to allocate the fixed asset investment of the whole project.

Table 10.7　　Results of investment allocation by storage capacity ratio method

Item	Investment Apportionment of shared projects			Investment in special project /10^4 yuan	Total investment in fixed assets /10^4 yuan	Proportion of investment in fixed assets /%
	Required storage capacity/$10^8 m^3$	Proportion of storage capacity/%	Allocation result /10^4 yuan			
Water supply	1.696	60.6	109539		109539	53.9
Flood control	0.68	24.3	43924		43924	21.6
Power generation	0.424	15.1	27294	22600	49894	24.5
Total	2.8	100	180757	22600	203357	100.0

The task of H Reservoir is water supply, flood control and power generation. According to the national economic evaluation results, the present value of water supply benefit is 2062.82 million yuan, the present value of flood control benefit is 6580600 million yuan, and the present value of power generation benefit is 683.27 million yuan, accounting for 60.6%, 19.3% and 20.1% respectively. According to this proportion, the fixed assets investment of the project is shared, and the water supply investment is 1232.35 million yuan, the flood control investment is 392.48 million yuan, and the power generation investment is 408.74 million yuan. See Table 10.8. From the result of investment allocation, the investment of generation function allocation is larger than that of special project, which shows that the allocation result of this method is reasonable.

Table 10.8　　Results of fixed-asset investment allocation by using benefit present value method

Item	Benefit present value/10^4 yuan	Proportion/%	Allocation result of fixed asset investment/10^4 yuan
Water supply	206282	60.6	123235
Flood control	65806	19.3	39248
Power generation	68327	20.1	40874
Total	340415	100.0	203357

3. Investment sharing results

Due to the limitations of various investment allocation methods, the average value of the above two methods is taken for this investment allocation. The fixed asset investment allocation of this project is: the water supply allocation investment is 1163.87 million yuan, the flood control allocation project investment is 415.86 million yuan, and the power generation allocation investment is 453.84 million-yuan, accounting for 57.2%, 20.5% and 22.3% respectively, as shown in Table 10.9.

Table 10.9 **Result of investment allocation**

Item	Allocation result of storage capacity proportion/10⁴ yuan	Allocation result of benefit present value /10⁴ yuan	Adopted value of investment allocation /10⁴ yuan	Proportion of investment in fixed assets/%
Water supply	109539	123235	116387	57.2
Flood control	43924	39248	41586	20.5
Power generation	49894	40874	45384	22.3
Total	203357	203357	203357	100.0

10.5.4 Total Cost Sharing

The total cost of the project includes repair cost, employee compensation, material cost, other expenses, management cost, insurance cost, water resource cost, reservoir fund, depreciation cost and financial cost, etc. The principle of total cost allocation is that the cost of services for a certain function is borne by the Department, and the cost of services for several departments is apportioned according to the proportion of fixed assets investment. If the water resources fee is required for water supply and power generation, the two departments shall bear their own requirements. The reservoir fund shall be paid by the power generation department and borne by the power generation. Other total costs shall be apportioned according to the fixed asset investment apportionment proportion, and the apportionment proportion of water supply, flood control and power generation is 57.2%, 20.5% and 22.3%, respectively. See Table 10.10 for the apportionment results of total cost before financing.

Table 10.10 **Result of total cost allocation before financing**   unit: 10⁴ yuan

Item	Water supply	Flood control	Power generation	Total
Repair fee	682	244	266	1192
Staff salary	371	133	145	648
Material fee	68	24	27	119
Other fees	112	40	44	196
Management fee	371	133	145	648
Insurance premium	170	61	66	298
Water resources fee	1092		128	1220
Reservoir fund			125	125
Depreciation cost	2792	1001	1088	4881
Total cost	5658	1636	2033	9327
Annual operation cost	2866	636	945	4446

10.5.5 Estimation of Cost Water Price and Cost Electricity Price

According to the results of total cost allocation, since there is no financial income from flood control, the annual operation cost allocated by flood control function is considered to be borne by water supply and power generation alone or shared by the cost proportion of both. The water price of water supply operation cost is 0.184 – 0.224 yuan/m³. The total water price of water supply cost is 0.363 – 0.403 yuan/m³. The electricity price of power generation operation cost is 0.060 – 0.101 yuan/m³, and the total water price of power generation cost is 0.130 – 0.171 yuan/m³. The total cost, annual operation cost, water price and electricity price are shown in Table 10.11.

Table 10.11　　　　　　　　　Result of total cost allocation

	Item	Scheme I Allocated cost of flood control is borne by water supply function	Scheme II Allocated cost of flood control is borne by power generation	Scheme III Allocated cost of flood control is borne by both water supply and power generation
Water supply	Total cost/10⁴ yuan	6294	5658	6116
	Annual operation cost/10⁴ yuan	3502	2866	3324
	Total cost water price/(yuan/m³)	0.403	0.363	0.392
	Operating cost water price/(yuan/m³)	0.224	0.184	0.213
Power generation	Total cost	2033	2669	2211
	Annual operation cost/10⁴ yuan	945	1581	1123
	Total cost price/[yuan/(kW·h)]	0.130	0.171	0.141
	Operating cost price/[yuan/(kW·h)]	0.060	0.101	0.072

10.6 Market Situation and User Affordability Analysis

10.6.1 Analysis of Water Supply Price

1. Current water prices

The current prices of tap water in the receiving area are four categories as follows: for civil water, the water price is 2 yuan/m³, and the corresponding sewage treatment fee is 0.85 yuan/m³. For special industries, the water price is 5 yuan/m³, and the corresponding sewage treatment fee is 1.50 yuan/m³. For other industries, the water price is 2.5 yuan/m³, and the corresponding sewage treatment fee is 1.10 yuan/m³. For comprehensive tap water, the water price is 2.3 yuan/m³, and the corresponding sewage treatment fee is 1.0 yuan/m³.

Currently, the receiving water area draws water from the urban rivers, and the raw water charge is 0.2 yuan/m³. The price of raw water for urban water supply near the receiving area is 0.2 - 0.9 yuan/m³.

2. Cost water price

The total cost water prices are different under different conditions. Bearing or partially bearing or even not bearing the cost of flood control operation, the total cost water prices of H Reservoir are 0.363 yuan/m³, 0.403 yuan/m³ and 0.392 yuan/m³, respectively. Correspondingly, the cost water prices of operation are 0.184 yuan/m³, 0.224 yuan/m³ and 0.213 yuan/m³, respectively.

3. Low - profit water price

The low - profit water price of the project is based on the cost water price to consider a certain investment profit rate. It can be calculated with the following equation:

the low - profit water price
= (total cost before finiancing of the water supply project
+ the fixed asset investment of water supply project × investment profit rate)
/ designed water supply amount / [1 − business tax rate
× (1 + sales tax surcharge rate)]

According to GSH [2007] No. 461 proposed by the State Taxation Administration, the water fee of water conservancy projects is included in the income obtained from providing natural water supply services to water users. The business tax shall be collected according to the tax items of "service industry". For service industry, the business tax rate is 5%. The sales tax surcharges include urban maintenance and construction tax, education surcharges and local education surcharges, and the tax rates are 5%, 3% and 2%, respectively. Therefore, the sales tax surcharge rate is 10%.

When the water supply does not need to bear the cost of flood control operation, the total cost of water supply is 0.363 yuan/m³. According to the fixed asset investment allocation, the fixed asset investment of water supply allocation project is 1163.87 million yuan. The water amount provided by the water supply project is 156 million m³. Given the investment profit rates of 1%, 2%, 3% and 4%, the calculated profit water prices are 0.464 yuan/m³, 0.542 yuan/m³, 0.621 yuan/m³ and 0.700 yuan/m³, respectively.

4. Profit water price

The apportioned investment of water supply of H Reservoir is 1163.87 million yuan. The capital and the loan amount should account for 35% and 65% of the total investment, respectively. In addition, the financial internal rates of return of capital should be 6%, 7% and 8%, respectively. Correspondingly, the estimated water supply prices, are 1.048 yuan/m³, 1.135 yuan/m³ and 1.229 yuan/m³, respectively.

5. User affordable water price

The designed water supply of the project is 156 million m³, with the amounts for domestic and industrial water supply are 68 million m³ and 88 million m³, respectively.

10.6 Market Situation and User Affordability Analysis

The designed level year of the project is 2030. The project is planned to take 4 years, with the start and completion in 2016 and 2019, respectively. Herein, we analyze the users' affordable water prices in 2020, 2025 and 2030.

(1) Affordable water price of industrial water users. The affordable water price of industrial water users can be analyzed according to the proportion of industrial water cost to industrial output value. According to *Regulation for Economic Evaluation of Water Conservancy Construction Projects* (*SL 72 -2013*), it is acceptable when the water cost of urban industrial enterprises accounts for 2%-5% of the industrial output value.

Herein, the industrial water cost accounts for 2% of the industrial output value as the expenditure standard of the water price that the industry can bear. According to the prediction results of water demand, the water demand of ten thousand yuan industrial added value in H Reservoir water supply area in 2030 is $26m^3/10000$ yuan, and the water demand of ten thousand yuan industrial output value is $6.5m^3/10^4$ yuan. According to the calculation of the industrial water cost accounting for 2% of the industrial output value, the water price that the industry can bear is 30.7 yuan/m^3. The affordable water prices that the industry can bear in 2025 and 2030 are 20.0 yuan/m^3 and 25.0 yuan/m^3, respectively, as shown in Table 10.12.

Table 10.12 The water price of industrial affordability in 2020, 2025 and 2030

Target year	Recycled water expenses of ten thousand yuan output value/yuan	Water consumption of ten thousand yuan output value/($m^3/10^4$ yuan)	Industrial bearable price /(yuan/m^3)
2020	200	10	20.0
2025	200	8	25.0
2030	200	6.5	30.7

Note: the cost of industrial water accounts for 2% of the industrial output value. This is taken as the standard of water fee for affordable water price.

(2) Affordable domestic water price for urban residents. The living affordable water price of urban residents can be analyzed according to the proportion of household or individual water resources fee to household income.

According to statistical data, the per capita disposable income of the residents in water receiving area of H Reservoir in 2014 was 25360 yuan. The actual annual water resources fee accounted for 0.3% of the distributable income.

According to *Regulation for Economic Evaluation of Water Conservancy Construction Projects* (*SL 72 -2013*), if the annual water resources fee of urban residents' accounts for 1.5%-3% of their annual disposable income, it is acceptable. Herein, the annual water resources fee accounts for 1.5% of annual disposable income as the living water resources fee standard of urban residents.

According to the historical growth of per capita disposable income of urban residents in the water receiving area as well as the prediction of 4% annual growth of per capita disposable income of urban residents, the per capita disposable income of urban residents in the water receiving area will be 47499 yuan, the quota of domestic water consumption of urban residents will be 160 L/(person · d) in 2030, and the proportion of water resources fee in the per capita disposable income of urban residents will be 1.5% as the analysis standard. As shown in Table 10.13, the affordable domestic water prices of urban residents in 2020, 2025 and 2030 are 9.4 yuan/m³, 10.7 yuan/m³, and 12.2 yuan/m³, respectively.

Table 10.13　　Analysis results of affordability of domestic water price for urban residents

Target year	Disposable income per capita /yuan	Annual water expenditure per capita/yuan	Annual water consumption per capita/(m³/year)	Affordable water price for residents/(yuan/m³)
2020	32088	481.3	51.1	9.4
2025	39040	585.6	54.8	10.7
2030	47499	712.5	58.4	12.2

Note: 1.5% of the annual disposable income of urban residents is taken as the standard of water resources fee.

(3) Urban comprehensive affordable water price. In 2030, the designed water supply of H Reservoir is 156 million m³, including 68 million m³ of domestic water supply and 88 million m³ of industrial water supply. The comprehensive affordable water price in the receiving area is 22.6 yuan/m³. As shown in Table 10.14, the affordable water price in 2020 and 2025 is 15.4 yuan/m³ and 18.8 yuan/m³, respectively.

Table 10.14　　Analysis results of affordability of urban comprehensive water price

Target year	Affordable water price for residents /(yuan/m³)	Affordable water price for industry /(yuan/m³)	Water supply amount/10⁸ m³			Comprehensive affordable water price for users /(yuan/m³)	Affordable water price at the outlet of H Reservoir /(yuan/m³)
			Industry	Living	Total		
2020	9.4	20.0	0.88	0.68	1.56	15.4	11.6
2025	10.7	25.0	0.88	0.68	1.56	18.8	15.0
2030	12.2	30.7	0.88	0.68	1.56	22.6	18.8

The affordable outlet water price of H Reservoir equals to the users' affordable price subtracting the low-profit water price of urban tap water (i.e., 2 yuan/m³) and the cost of water conveyance project (i.e., 1.8 yuan/m³, including 1.0 yuan/m³ for the water conveyance project, and 0.8 yuan/m³ for the sewage treatment). In 2030, the affordable outlet water price of the H Reservoir is 18.8 yuan/m³.

10.6.2　Analysis of On-grid Price

1. Current on-grid Price

For the newly-built small hydropower stations (10MW ⩽ installed capacity < 50

MW, 8%≤capacity coefficient<20%) on provincial grid, the on-grid benchmark price including and excluding VAT is 0.388 yuan/(kW·h) and 0.332 yuan/(kW·h), respectively. These are given based on the power grid data of H Reservoir power station.

The electricity price excluding VAT, recently approved by the Provincial Price Bureau, of large and medium-sized reservoir power stations is up to 0.357 yuan/(kW·h).

2. Cost price

The cost prices are different under different conditions. Specifically, when the power generation bears or partially bears or even does not bear the cost of flood control operation, the total cost prices are 0.130 yuan/(kW·h) and 0.171 yuan/(kW·h), 0.141 yuan/(kW·h), respectively. And the corresponding operation cost prices are 0.060 yuan/(kW·h), 0.101 yuan/(kW·h), and 0.072 yuan/(kW·h), respectively.

3. On-grid price meeting the requirement of financial internal rate of return of capital

The allocation investment of H Reservoir power generation is 453.84 million yuan. According to the capital accounts for 20% of the total investment and the loan accounts for 80% of the total investment, it is estimated that the on-grid price meeting the requirement of 8% of the financial internal rate of return of capital is 0.341 yuan/(kW·h).

10.7 Measurement of Loaning Ability

10.7.1 Calculation Principle

Based on the proposed water supply price, on-grid price and loan repayment period, the current long-term loan interest rate is adopted in the measurement of loaning ability. It is for the calculation of the maximum loaning ability and the required capital amount of each scheme and the identification of fund-raising schemes, under the current fiscal and taxation system and bank credit conditions. No profit shall be distributed on all capital during the loan period.

10.7.2 Scheme Design

1. Scheme of Water Price

The water price of water supply cost of H Reservoir is 0.363 yuan/m³. The calculated profit water prices, given the investment profit rates of 1%, 2%, 3% and 4%, are 0.464 yuan/m³, 0.542 yuan/m³, 0.621 yuan/m³ and 0.700 yuan/m³, respectively. The profit water price of 6%, 7% and 8% of the capital fund of water supply project is 1.048 yuan/m³, 1.135 yuan/m³ and 1.229 yuan/m³. The affordable water price in 2020 would be 4.4-11.8 yuan/m³, whereas it would be 7.6-18.2 yuan/m³ in 2030. Six schemes for water price are proposed, including 0.36 yuan/m³, 0.55 yuan/m³, 0.70 yuan/m³, 0.85 yuan/m³, 1.05 yuan/m³ and 1.23 yuan/m³.

2. Scheme of on-grid price

H Reservoir is an annual regulating reservoir. The installed capacity is 60 MW, and the storage coefficient is 17.3%. The benchmark on-grid price for small hydropower stations is 0.332 yuan/(kW·h), and the maximum on-grid price for large and medium-sized reservoir power stations is 0.357 yuan/(kW·h). Considering these two prices, four schemes for on-grid price are proposed: 0.332 yuan/(kW·h), 0.357 yuan/(kW·h), 0.38 yuan/(kW·h) and 0.40 yuan/(kW·h).

10.7.3 Financial Income Estimation

The financial income of the project mainly includes the water fee income and power generation income of urban industry and residents.

1. Water supply income

$$Water\ supply\ income = water\ supply\ quantity \times water\ price$$

The designed annual water supply of H Reservoir is 15600 m^3, and the initial water supply takes up 30%. The benefits of designed water supply will be achieved in the 11th year.

2. Generation revenue

$$Generation\ revenue = on\text{-}grid\ energy \times on\text{-}grid\ price$$

Three mixed-flow turbine generator units are arranged in the power plant. The total installed capacity is 60 MW. The multi-year average electricity output and on-grid energy is 160 million kW·h, and 156.48 million kW·h, respectively.

10.7.4 Taxes

The taxes paid for this project include Value added tax (VAT), business tax, sales tax surcharges and income tax.

1. Value added tax (VAT)

In the water receiving area of this project, VAT is charged for water supply. The VAT rate is 13%, the same as that of tap water supply. According to *The Provisional Regulations of the People's Republic of China on Value Added Tax (2009)*:

$$the\ output\ tax\ of\ electric\ power\ products = sales\ amount \times tax\ rate$$

The VAT rate of the hydropower stations (installed capacity ≥ 50 MW) is 17%. Herein, the VAT rate of the power generation income of this project is 17%.

2. Business tax

According to *The Notice of the Ministry of Finance and the State Development Planning Commission on Converting Some Administrative and Institutional Charges into Business Service Charges (Prices) (CZ [2001] No. 94)*, the water charges of water conservancy projects are changed from administrative and institutional charges to business service charges. The State Administration of Taxation clearly pointed out in National Tax

10.7 Measurement of Loaning Ability

Letter [2007] No. 461 that the water charges of water conservancy projects collected from users by water conservancy engineering units belong to the category of water charges collected from users. According to the current turnover tax policy, the income from the natural water supply service is not subject to value-added tax, but to business tax according to the tax items of "service industry". The business tax rate of "service industry" is 5%.

3. Sales tax surcharges

Sales tax surcharges include urban maintenance & construction tax, educational surtax and local educational surtax. These are taxes included in the calculated prices.

According to *Interim Regulations of the People's Republic of China on Urban Maintenance & Construction Tax* (GF [1985] No. 19), to strengthen the maintenance and construction of cities, expand and stabilize the fund sources of urban maintenance and construction, the urban maintenance & construction tax is levied. The urban maintenance & construction tax is calculated on the basis of the taxpayers-paid product tax, VAT and business tax. Correspondingly, they shall be simultaneously paid with the product tax, VAT, and business tax. According to different conditions, the urban maintenance & construction tax rates are set as follows: If the taxpayer is located in the urban area, the tax rate is 7%; If the taxpayer is located in the county or town, the tax rate is 5%; If the taxpayer is not located in the urban area, county or town, the tax rate is 1%. The taxpayer of this project is located in the county seat, and the urban maintenance and construction tax rate is 5%, based on VAT and business tax.

According to *Temporary Stipulation on Collecting Educational Extra-fee* (issued by the State Council on April 28, 1986), and to implement *Decision on Educational System Reform by the CPCCC*, expand the sources of local education funds, collect educational surtax. The educational surtax is based on the VAT, sales tax and consumption tax paid by various units and individuals. The rate of educational surtax is 3%, and should be paid together with VAT, business tax and consumption tax. The education surcharge tax rate for this project is 5%, based on VAT and business tax.

According to "*Circular of the Ministry of Finance on Issues Related to the Unification of Local Education Additional Policies*" (CZ [2010] No. 98), in order to implement the *Outline of the National Medium and Long-term Education Reform and Development Plan (2010-2020)*, further standardize and expand the sources of financial education funds, support the development of local education, the local education additional collection standards are unified. The collection standard of local educational surtax shall be 2% of VAT, business tax and consumption tax actually paid by units and individuals (including enterprises with foreign investment, foreign enterprises and foreign individuals). For provinces that have been approved by the Ministry of Finance and whose collection standard is less than 2%, the collection standard of local education surtax should be adjusted to 2%. The education surcharge tax rate for this project is 5%, based on VAT and business tax.

4. Income tax

The income tax is calculated based on the amount of taxable income. The amount of taxable income of this project is the balance of the sales revenue of power generation deducting the cost and sales tax. According to the newly promulgated *Enterprise Income Tax Law of the People's Republic of China*, the income tax is levied at 25%.

10.7.5 Estimation Conditions

1. Calculation period

The construction, operation and calculation periods of H Reservoir are 4 years, 40 years, and 44 years, respectively.

2. Fixed assets investment

The fixed asset investment of H Reservoir is 2033.57 million yuan. The annual investment is shown in Table 10.2.

3. Total cost

See Section 10.3.2 for the total cost parameter values of the project.

4. Credit period and loan rate

The project is a large-scale infrastructure project. There are two schemes for the credit period, i.e., 20 years and 25 years. The loan interest rate is current five-year long-term loan interest rate at 5.65%. Herein, the interest is calculated once a year.

The capital to repay the loan principal is undistributed profit and depreciation expense.

10.7.6 Calculation Results of Loan Capacity

According to above schemes and conditions, the loan capacity is calculated according to the short-term loan period of no more than 5 years and no profit is paid during the loan repayment period. See Table 10.15 for the estimation results of loan capacity. The relationship

Table 10.15   Calculation results of H Reservoir loan capacity

Serial number	Loan period /a	On-grid energy /[yuan /(kW·h)]	Water supply price/(yuan /m³)	Loan capacity /10⁴ yuan	Capital fund /10⁴ yuan	Interest during construction period /10⁴ yuan	Total investment /10⁴ yuan	FIRR of project investment /%	FIRR of capital fund /%	Proportion of loan principal to construction investment /%
1	20	0.332	0.36	33266	170091	4172	207529	0.25	−0.41	16.36
2			0.55	45065	158292	5638	208995	1.97	1.01	22.16
3			0.70	54384	148973	6790	210147	3.07	1.97	26.74
4			0.85	63380	139977	7897	211254	4.03	2.87	31.17
5			1.05	74926	128431	9313	212670	5.16	3.98	36.84
6			1.23	85106	118251	10555	213912	6.07	4.96	41.85

10.7 Measurement of Loaning Ability

Continued

Serial number	Loan period /a	On-grid energy /[yuan/(kW·h)]	Water supply price/(yuan/m³)	Loan capacity /10⁴ yuan	Capital fund /10⁴ yuan	Interest during construction period /10⁴ yuan	Total investment /10⁴ yuan	FIRR of project investment /%	FIRR of capital fund /%	Proportion of loan principal to construction investment /%
7		0.357	0.36	36738	166619	4604	207961	0.54	−0.19	18.07
8			0.55	48530	154827	6066	209423	2.20	1.19	23.86
9			0.70	57850	145507	7217	210574	3.28	2.15	28.45
10			0.85	66766	136591	8313	211670	4.22	3.04	32.83
11			1.05	78239	125118	9718	213075	5.33	4.15	38.47
12			1.23	88319	115038	10947	214304	6.22	5.13	43.43
13	20	0.38	0.36	39941	163416	5002	208359	0.80	0.00	19.64
14			0.55	51741	151616	6463	209820	2.41	1.36	25.44
15			0.70	60981	142376	7602	210959	3.46	2.31	29.99
16			0.85	69851	133506	8692	212049	4.39	3.19	34.35
17			1.05	81267	122090	10087	213444	5.48	4.31	39.96
18			1.23	91261	112096	11304	214661	6.36	5.29	44.88
19		0.40	0.36	42718	160639	5347	208704	1.02	0.17	21.01
20			0.55	54519	148838	6806	210163	2.59	1.51	26.81
21			0.70	63673	139684	7933	211290	3.62	2.45	31.31
22			0.85	72493	130864	9015	212372	4.53	3.33	35.65
23			1.05	83905	119452	10409	213766	5.61	4.45	41.26
24			1.23	93828	109529	11616	214973	6.49	5.44	46.14
25	25	0.332	0.36	38840	164517	4866	208223	0.24	−0.67	19.10
26			0.55	52608	150749	6570	209927	1.96	0.77	25.87
27			0.70	63497	139860	7912	211269	3.06	1.76	31.22
28			0.85	74371	128986	9245	212602	4.02	2.69	36.57
29			1.05	88480	114877	10966	214323	5.15	3.89	43.51
30			1.23	100882	102475	12471	215828	6.05	4.98	49.61
31		0.357	0.36	42889	160468	5368	208725	0.53	−0.47	21.09
32			0.55	56668	146689	7071	210428	2.19	0.95	27.87
33			0.70	67551	135806	8409	211766	3.27	1.93	33.22
34			0.85	88438	124919	9742	213099	4.21	2.87	38.57
35			1.05	92453	110904	11449	214806	5.32	4.08	45.46
36			1.23	104789	98568	12944	216301	6.21	5.18	51.53

Continued

Serial number	Loan period /a	On-grid energy /[yuan/(kW·h)]	Water supply price/(yuan/m³)	Loan capacity /10⁴ yuan	Capital fund /10⁴ yuan	Interest during construction period /10⁴ yuan	Total investment /10⁴ yuan	FIRR of project investment /%	FIRR of capital fund /%	Proportion of loan principal to construction investment /%
37		0.38	0.36	46633	156724	5832	209189	0.79	−0.29	22.93
38			0.55	60396	142961	7530	210887	2.40	1.11	29.70
39			0.70	71290	132067	8868	212225	3.45	2.09	35.06
40			0.85	82145	121212	10195	213552	4.38	3.03	40.39
41			1.05	96139	107218	11896	215253	5.47	4.25	47.28
42	25		1.23	108384	94973	13378	216735	6.35	5.37	53.30
43		0.40	0.36	49874	153483	6233	209590	1.01	−0.13	24.53
44			0.55	63653	139704	7931	211288	2.58	1.25	31.30
45			0.70	74530	128827	9264	212621	3.61	2.23	36.65
46			0.85	85327	118030	10582	213939	4.52	3.17	41.96
47			1.05	99312	104045	12281	215638	5.60	4.41	48.84
48			1.23	111498	91859	13753	217110	6.47	5.54	54.83

between water supply price, on-grid electricity price and loan principal is shown in Figure 10.1 and Figure 10.2.

From Table 10.15, Figure 10.1 and Figure 10.2, the water supply price has a great

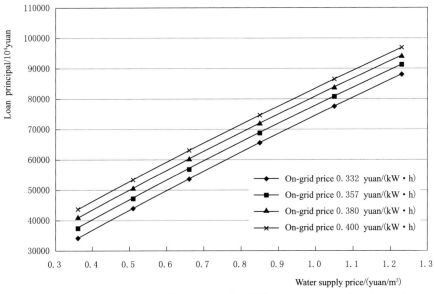

Figure 10.1 Loan principal relation under different water supply price and on-grid prices of Reservoir H (repayment period=20 years)

impact on the maximum loan amount of the project. As the project has a long production period, the loan amount is controlled by the short-term loan period of no more than 5 years at the initial operation stage, so the loan period has little impact on the loan principal of the project.

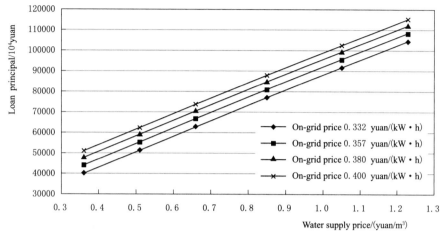

Figure 10.2 Loan principal relation under different water supply price and on-grid prices of Reservoir H (repayment period=25 years)

10.8 Financing Schemes

1. Recommended on-grid price

Currently, the benchmark electricity price in this province for small hydropower stations is 0.332 yuan/(kW·h). And the maximum electricity price for large and medium-sized reservoir power stations is 0.357 yuan/(kW·h). As a kind of clean energy, the hydropower can be used in peak load regulation in the power grid with strong competitiveness. After the project completion, it will take 5-6 years before online water supply. Considering the increasing trend, the recommended electricity price is 0.38 yuan/(kW·h).

2. Recommended water supply price

Water in the receiving area is drawn from the urban river. The raw water cost is 0.2 yuan/m^3. The total cost water price of H Reservoir is 0.363 yuan/m^3. The operation cost water price is 0.184 yuan/m^3. The profit water price calculated according to the profit margin of 4% of the shared investment in water supply is 0.700 yuan/m^3. The acceptable water price at the end of the project is 18.8 yuan/m^3.

The water transportation project is also necessary to supply water to the receiving area. According to data analysis, the profit water price of the water transportation project, between the end of water supply to the urban waterworks, is 1.0 yuan/m^3.

The price of raw water for urban water supply near the receiving area is 0.2-0.9 yuan/m^3.

The current water supply price in the receiving area is relatively low, so there is some space to increase the water supply price. In addition, considering the psychological endurance of water users, at this stage of this project, the water supply price is recommended to be 0.70 yuan/m³. Correspondingly, the investment profit rate is 4%, and the raw water price at the inlet of the urban water plant is 1.70 yuan/m³. Considering of the cost of tap water and the low profit, the water price to the users is about 4.5 yuan/m³.

3. Financing Schemes

Given the water supply price as 0.70 yuan/m³, the on-grid price as 0.38 yuan/(kW·h), and the loan repayment period as 25 years. The principal of the project loan is 712.90 million yuan. The interest during the construction period is 88.68 million yuan. The required capital for this project is 1320.67 million yuan.

For this project, there are still many profits after the loan repayment. In order to reduce the financial capital pressure, parts of the enterprise capital can be taken into account. According to the requirements of 6%-10% as the financial internal rate of return (FIRR) of the enterprise capital, the amount of the enterprise capital that can be absorbed is estimated to be 497.89 - 215.80 million yuan, as shown in Table 10.16.

Table 10.16   Absorbable enterprise capital amount of H Reservoir

Item	FIRR of enterprise capital, 6%	FIRR of enterprise capital, 7%	FIRR of enterprise capital, 8%	FIRR of enterprise capital, 9%	FIRR of enterprise capital, 10%
Loan period/a	25				
On-grid price/[yuan/(kW·h)]	0.38				
Water supply price/(yuan/m³)	0.70				
Loan principal/10⁴ yuan	71290				
Capital fund/10⁴ yuan	132067				
Interest during construction period/10⁴ yuan	8868				
Total investment/10⁴ yuan	212225				
FIRR of project investment	3.45%				
FIRR of project capital fund	2.09%				
Proportion of loan principal to fixed-asset investment	35.06%				
Absorbable enterprise capital amount/10⁴ yuan	49789	39818	32158	26229	21580
Amount of government capital/10⁴ yuan	82278	92249	99909	105839	110488
Proportion of enterprise capital to capital fund/%	37.70	30.15	24.35	19.86	16.34

The FIRR of enterprise capital is required to be 8%. Accordingly, the absorbable enterprise capital is 321.58 million yuan, and the needed amount of government capital is 999.09 million yuan.

The estimate of the loan capacity of this project is controlled by short-term loan for no more than 5 years. Therefore, the calculated enterprise capital has no profit in the first 6 years (5th to 10th year) of the operation period. The average profit during the loan repayment period is 20.34 million yuan (the 5th to the 25th year), and the average rate of profit payable is 6.4%. If the investors want to improve the profit payable during the loan repayment period, the amount of enterprise capital needs to be increased.

10.9 Financial Evaluation

10.9.1 Calculation Condition

1. Financing plan

According to the recommended financing plan, the loan principal, the enterprise capital and the government capital of this project are 712.90 million yuan, 321.58 million yuan, and 999.09 million yuan, respectively. Additionally, the recommended water price, the on-grid price, and the loan repayment period are 0.70 yuan/m^3, 0.38 yuan/(kW·h), and 25 years, respectively.

The long-term loan interest rate is 5.65%. The loan period is 25 years. The interest during the construction period is 88.68 million yuan. The total project investment is 2122.25 million yuan. The investment plan and financing plan of this project are provided in Table 10.17.

Total 10.17　　　　Project investment plan and fund raising　　　　unit: 10^4 yuan

Serial number	Item	Total	1	2	3	4	5
1	Total investment	212225	49247	58845	55742	48391	
1.1	Investment in fixed assets	203357	48736	57212	52920	44489	
1.2	Interest during construction period	8868	511	1633	2822	3902	
2	Working capital	445					445
3	Fund raising	212670	49247	58845	55742	48391	445
3.1	Capital fund	132201	30646	36619	34688	30114	134
3.1.1	For fixed-asset investment	132607	30646	36619	34688	30114	

Continued

Serial number	Item	Total	1	2	3	4	5
	Government investment	99909	23184	27702	26242	22781	
	Enterprise capital fund	32158	7462	8917	8447	7333	
3.1.2	For working capital	134					134
	Government investment						
	Enterprise capital fund	134					134
3.2	Loan capital	80469	18601	22226	21854	18277	312
3.2.1	For fixed-asset investment	71290	18090	20593	18232	14375	
3.2.2	For interest during construction period	8868	511	1633	2822	3902	
3.2.3	For working capital	312					312
3.3	Other capital						

2. Total cost

The costs of this project include repair cost (12.81 million yuan), employee compensation (6.48 million yuan), material cost (1.28 million yuan), other expenses (2.06 million yuan), management cost (6.48 million yuan), fixed asset insurance cost (3.2 million yuan), water resources cost (12.2 million yuan, including 10.92 million yuan for water supply, 1.28 million yuan for power generation), reservoir fund (1.25 million yuan), depreciation (50.91 million yuan) and financial cost (45.45 million − 0.16 million yuan). To sum up, the total cost is 134.86 million − 96.85 million yuan, and the annual operation cost is 45.76 million yuan. All details can be seen in Table 10.18.

3. Project financial income

The financial income of this project mainly includes the water supply income and power generation income collected from urban industries and residents. In normal years, it is 168.66 million yuan, including 109.20 million yuan of annual water supply income and 59.46 million yuan of annual power generation income. The project profit and profit distribution are provided in Table 10.19.

10.9.2 Financial Evaluation

The construction and operation periods of H Reservoir are 4 years and 40 years, respectively. And the calculation period is 44 years. Results of the financial evaluation are provided in Table 10.20.

10.9 Financial Evaluation

Table 10.18 Total cost estimation

unit: 10^4 yuan

Serial number	Item	Total	1	2	3	4	5	6	7	8	9	10	11	12	13	14	15	16	17	18	19	20	21
1	Total cost	426372	0	0	0	0	8921	9643	10366	11088	11810	12533	13094	13006	12909	12803	12686	12482	12266	12038	11798	11543	11274
2	Depreciation expense	203736	0	0	0	0	5093	5093	5093	5093	5093	5093	5093	5093	5093	5093	5093	5093	5093	5093	5093	5093	5093
3	Actual depreciation cost	187598	0	0	0	0	565	1251	1973	2732	3531	4371	5093	5093	5093	5093	5093	5093	5093	5093	5093	5093	5093
4	Amortization fee	0	0	0	0	0	0	0	0	0	0	0	0	0	0	0	0	0	0	0	0	0	0
5	Financial cost	59944	0	0	0	0	4545	4504	4429	4315	4162	3968	3730	3566	3393	3210	3017	2813	2597	2369	2128	1874	1605
5.1	Long term loan interest expense	58748	0	0	0	0	4529	4411	4287	4155	4016	3869	3714	3550	3377	3194	3001	2797	2581	2353	2112	1858	1589
5.2	Working capital interest expenditure	635	0	0	0	0	16	16	16	16	16	16	16	16	16	16	16	16	16	16	16	16	16
5.3	Other short-term loan interest expenditure	561	0	0	0	0	0	78	126	144	130	83	0	0	0	0	0	0	0	0	0	0	0
6	Annual operation cost	178830	0	0	0	0	3811	3888	3964	4041	4117	4194	4270	4347	4423	4499	4576	4576	4576	4576	4576	4576	4576
6.1	Staff salary	25920	0	0	0	0	648	648	648	648	648	648	648	648	648	648	648	648	648	648	648	648	648
6.2	Repair fee	51229	0	0	0	0	1281	1281	1281	1281	1281	1281	1281	1281	1281	1281	1281	1281	1281	1281	1281	1281	1281
6.3	Reservoir fund	5008	0	0	0	0	125	125	125	125	125	125	125	125	125	125	125	125	125	125	125	125	125
6.4	Material fee	5123	0	0	0	0	128	128	128	128	128	128	128	128	128	128	128	128	128	128	128	128	128
6.5	Water resources fee	44596	0	0	0	0	456	532	608	685	761	838	914	991	1067	1144	1220	1220	1220	1220	1220	1220	1220
6.6	Fixed-asset insurance premium	12807	0	0	0	0	320	320	320	320	320	320	320	320	320	320	320	320	320	320	320	320	320
6.7	Other costs	8227	0	0	0	0	206	206	206	206	206	206	206	206	206	206	206	206	206	206	206	206	206
6.8	Management fee	25920	0	0	0	0	648	648	648	648	648	648	648	648	648	648	648	648	648	648	648	648	648

10 Case Study of Economic Evaluation of Water Conservancy Projects

Continued

Serial number	Item	22	23	24	25	26	27	28	29	30	31	32	33	34	35	36	37	38	39	40	41	42	43	44
1	Total cost	10991	10691	10374	10039	9685	9685	9685	9685	9685	9685	9685	9685	9685	9685	9685	9685	9685	9685	9685	9685	9685	9685	9685
2	Depreciation expense	5093	5093	5093	5093	5093	5093	5093	5093	5093	5093	5093	5093	5093	5093	5093	5093	5093	5093	5093	5093	5093	5093	5093
3	Actual depreciation	5093	5093	5093	5093	5093	5093	5093	5093	5093	5093	5093	5093	5093	5093	5093	5093	5093	5093	5093	5093	5093	5093	5093
4	Amortization expense	0	0	0	0	0	0	0	0	0	0	0	0	0	0	0	0	0	0	0	0	0	0	0
5	Financial cost	1321	1021	704	370	16	16	16	16	16	16	16	16	16	16	16	16	16	16	16	16	16	16	16
5.1	Long-term loan interest expenditure	1305	1005	689	354	0	0	0	0	0	0	0	0	0	0	0	0	0	0	0	0	0	0	0
5.2	Interest expense of working capital loan	16	16	16	16	16	16	16	16	16	16	16	16	16	16	16	16	16	16	16	16	16	16	16
5.3	Other short-term loan interest expenditure	0	0	0	0	0	0	0	0	0	0	0	0	0	0	0	0	0	0	0	0	0	0	0
6	Annual operation cost	4576	4576	4576	4576	4576	4576	4576	4576	4576	4576	4576	4576	4576	4576	4576	4576	4576	4576	4576	4576	4576	4576	4576
6.1	Staff salary	648	648	648	648	648	648	648	648	648	648	648	648	648	648	648	648	648	648	648	648	648	648	648
6.2	Management fee	1281	1281	1281	1281	1281	1281	1281	1281	1281	1281	1281	1281	1281	1281	1281	1281	1281	1281	1281	1281	1281	1281	1281
6.3	Reservoir fund	125	125	125	125	125	125	125	125	125	125	125	125	125	125	125	125	125	125	125	125	125	125	125
6.4	Material fee	128	128	128	128	128	128	128	128	128	128	128	128	128	128	128	128	128	128	128	128	128	128	128
6.5	Water resources fee	1220	1220	1220	1220	1220	1220	1220	1220	1220	1220	1220	1220	1220	1220	1220	1220	1220	1220	1220	1220	1220	1220	1220
6.6	Fixed-asset insurance premium	320	320	320	320	320	320	320	320	320	320	320	320	320	320	320	320	320	320	320	320	320	320	320
6.7	Other cost	206	206	206	206	206	206	206	206	206	206	206	206	206	206	206	206	206	206	206	206	206	206	206
6.8	Management fee	648	648	648	648	648	648	648	648	648	648	648	648	648	648	648	648	648	648	648	648	648	648	648

10.9 Financial Evaluation

Table 10.19 Profit and profit distribution

unit: 10^4 yuan

Serial number	Item	Total	1	2	3	4	5	6	7	8	9	10	11	12	13	14	15	16	17	18	19	20	21
1	Sales income	632617	0	0	0	0	9222	9987	10751	11516	12280	13044	13809	14573	15338	16102	16866	16866	16866	16866	16866	16866	16866
1.1	Power generation income	237859	0	0	0	0	5946	5946	5946	5946	5946	5946	5946	5946	5946	5946	5946	5946	5946	5946	5946	5946	5946
	On-grid energy/(10^8 kW·h)	62.595	0	0	0	0	1.565	1.565	1.565	1.565	1.565	1.565	1.565	1.565	1.565	1.565	1.565	1.565	1.565	1.565	1.565	1.565	1.565
	On-grid price/[yuan/(kW·h)]			0	0	0	0.38	0.38	0.38	0.38	0.38	0.38	0.38	0.38	0.38	0.38	0.38	0.38	0.38	0.38	0.38	0.38	0.38
1.2	Urban water supply income	394758	0	0	0	0	3276	4040	4805	5569	6334	7098	7862	8627	9391	10156	10920	10920	10920	10920	10920	10920	10920
	Urban water supply amount/10^8 m^3	56	0	0	0	0	0.4680	0.5772	0.6864	0.7956	0.9048	1.0140	1.1232	1.2324	1.3416	1.4508	1.5600	1.5600	1.5600	1.5600	1.5600	1.5600	1.5600
	Water supply price/(yuan/m^3)	28	0	0	0	0	0.700	0.700	0.700	0.700	0.700	0.700	0.700	0.700	0.700	0.700	0.700	0.700	0.700	0.700	0.700	0.700	0.700
2	Subsidy income	0	0	0	0	0	0	0	0	0	0	0	0	0	0	0	0	0	0	0	0	0	0
3	Sales tax and surcharges	26564	0	0	0	0	301	344	386	428	470	512	554	596	638	680	722	722	722	722	722	722	722
4	Total cost	426372	0	0	0	0	8921	9643	10366	11088	11810	12533	13094	13006	12909	12803	12686	12482	12266	12038	11798	11543	11274
5	Total profit	179681	0	0	0	0	0	0	0	0	0	0	162	971	1790	2619	3458	3663	3878	4106	4347	4601	4870
6	Make up for the loss of previous year	0																					
7	Taxable income	179681	0	0	0	0	0	0	0	0	0	0	162	971	1790	2619	3458	3663	3878	4106	4347	4601	4870
8	Income tax	44920	0	0	0	0	0	0	0	0	0	0	40	243	448	655	865	916	970	1027	1087	1150	1218
9	Profit after tax	134761	0	0	0	0	0	0	0	0	0	0	121	729	1343	1964	2594	2747	2909	3080	3260	3451	3653
10	Undistributed profit at the beginning of the period	0																					

10 Case Study of Economic Evaluation of Water Conservancy Projects

Continued

Serial number	Item	Total	1	2	3	4	5	6	7	8	9	10	11	12	13	14	15	16	17	18	19	20	21
11	Profit available for distribution	134761	0	0	0	0	0	0	0	0	0	0	121	729	1343	1964	2594	2747	2909	3080	3260	3451	3653
12	Withdrawal of statutory surplus reserve	13476	0	0	0	0	0	0	0	0	0	0	12	73	134	196	259	275	291	308	326	345	365
13	Distributable profit	121285	0	0	0	0	0	0	0	0	0	0	109	656	1209	1768	2334	2472	2618	2772	2934	3106	3287
14	Profit distribution among investors	0	0	0	0	0	0	0	0	0	0	0	0	0	0	0	0	0	0	0	0	0	0
15	Undistributed profit	121285	0	0	0	0	0	0	0	0	0	0	109	656	1209	1768	2334	2472	2618	2772	2934	3106	3287
16	Profit before interest and tax	239625	0	0	0	0	4545	4504	4429	4315	4162	3968	3892	4538	5183	5829	6475	6475	6475	6475	6475	6475	6475
17	Before interest, tax, depreciation and amortization	427223	0	0	0	0	5110	5755	6401	7047	7693	8339	8985	9631	10277	10923	11569	11569	11569	11569	11569	11569	11569

Serial number	Item	22	23	24	25	26	27	28	29	30	31	32	33	34	35	36	37	38	39	40	41	42	43	44
1	Sales income	16866	16866	16866	16866	16866	16866	16866	16866	16866	16866	16866	16866	16866	16866	16866	16866	16866	16866	16866	16866	16866	16866	16866
1.1	Power generation sales revenue	5946	5946	5946	5946	5946	5946	5946	5946	5946	5946	5946	5946	5946	5946	5946	5946	5946	5946	5946	5946	5946	5946	5946
	On-grid energy/(10^8 kW·h)	1.565	1.565	1.565	1.565	1.565	1.565	1.565	1.565	1.565	1.565	1.565	1.565	1.565	1.565	1.565	1.565	1.565	1.565	1.565	1.565	1.565	1.565	1.565
	On-grid price/[yuan/(kW·h)]	0.38	0.38	0.38	0.38	0.38	0.38	0.38	0.38	0.38	0.38	0.38	0.38	0.38	0.38	0.38	0.38	0.38	0.38	0.38	0.38	0.38	0.38	0.38
1.2	Urban water supply income	10920	10920	10920	10920	10920	10920	10920	10920	10920	10920	10920	10920	10920	10920	10920	10920	10920	10920	10920	10920	10920	10920	10920
	Urban water supply amount/$10^8 m^3$	1.5600	1.5600	1.5600	1.5600	1.5600	1.5600	1.5600	1.5600	1.5600	1.5600	1.5600	1.5600	1.5600	1.5600	1.5600	1.5600	1.5600	1.5600	1.5600	1.5600	1.5600	1.5600	1.5600
	Water supply price/(yuan/m^3)	0.700	0.700	0.700	0.700	0.700	0.700	0.700	0.700	0.700	0.700	0.700	0.700	0.700	0.700	0.700	0.700	0.700	0.700	0.700	0.700	0.700	0.700	0.700

10.9 Financial Evaluation

Continued

Serial number	Item	22	23	24	25	26	27	28	29	30	31	32	33	34	35	36	37	38	39	40	41	42	43	44
2	Subsidy income	0	0	0	0	0	0	0	0	0	0	0	0	0	0	0	0	0	0	0	0	0	0	0
3	Sales income and surcharges	722	722	722	722	722	722	722	722	722	722	722	722	722	722	722	722	722	722	722	722	722	722	722
4	Total cost	10991	10691	10374	10039	9685	9685	9685	9685	9685	9685	9685	9685	9685	9685	9685	9685	9685	9685	9685	9685	9685	9685	9685
5	Total profit	5154	5454	5771	6106	6459	6459	6459	6459	6459	6459	6459	6459	6459	6459	6459	6459	6459	6459	6459	6459	6459	6459	6459
6	Make up for the loss of the previous year																							
7	Taxable income	5154	5454	5771	6106	6459	6459	6459	6459	6459	6459	6459	6459	6459	6459	6459	6459	6459	6459	6459	6459	6459	6459	6459
8	Income tax	1289	1363	1443	1526	1615	1615	1615	1615	1615	1615	1615	1615	1615	1615	1615	1615	1615	1615	1615	1615	1615	1615	1615
9	Profit after tax	3866	4090	4328	4579	4845	4845	4845	4845	4845	4845	4845	4845	4845	4845	4845	4845	4845	4845	4845	4845	4845	4845	4845
10	Undistributed profit at the beginning of the period																							
11	Profit available for distribution	3866	4090	4328	4579	4845	4845	4845	4845	4845	4845	4845	4845	4845	4845	4845	4845	4845	4845	4845	4845	4845	4845	4845
12	Withdrawal of statutory surplus reserve	387	409	433	458	484	484	484	484	484	484	484	484	484	484	484	484	484	484	484	484	484	484	484
13	Profit available for distribution	3479	3681	3895	4121	4360	4360	4360	4360	4360	4360	4360	4360	4360	4360	4360	4360	4360	4360	4360	4360	4360	4360	4360
14	Profit distribution among investors	0	0	0	0	0	0	0	0	0	0	0	0	0	0	0	0	0	0	0	0	0	0	0
15	Undistributed profit	3479	3681	3895	4121	4360	4360	4360	4360	4360	4360	4360	4360	4360	4360	4360	4360	4360	4360	4360	4360	4360	4360	4360
16	Profit before interest and tax	6475	6475	6475	6475	6475	6475	6475	6475	6475	6475	6475	6475	6475	6475	6475	6475	6475	6475	6475	6475	6475	6475	6475
17	Before interest, tax, depreciation and amortization	11569	11569	11569	11569	11569	11569	11569	11569	11569	11569	11569	11569	11569	11569	11569	11569	11569	11569	11569	11569	11569	11569	11569

10 Case Study of Economic Evaluation of Water Conservancy Projects

Table 10.20　　　　　　　　　Summary of financial indicators

Serial number	Item	Unit	Indicators	Note
1	Total investment	10^4 yuan	212225	
1.1	Fixed-asset investment	10^4 yuan	203357	
1.2	Interest during construction period	10^4 yuan	8768	
2	Working capital	10^4 yuan	445	
3	Price			
3.1	On-grid price	yuan/(kW·h)	0.380	Excluding value added tax
3.2	Urban water supply price	yuan/m³	0.700	Excluding value added tax
4	Total amount of sales income	10^4 yuan	632617	
5	Total cost	10^4 yuan	426372	
6	Total sales tax and surcharges	10^4 yuan	26564	Including business tax
7	Total profits	10^4 yuan	179681	
8	Profit indicators			
8.1	Return on investment, ROI	%	2.82	
8.2	Return on equity investment, ROE	%	2.55	
8.3	FIRR of project investment (before income tax)	%	3.45	
8.4	FIRR of project investment (after income tax)	%	2.66	
8.5	FNPV of project investment (before income tax)	10^4 yuan	−88351	
8.6	FNPV of project investment (after income tax)	10^4 yuan	−100088	
8.7	Payback time (before income tax)	years	24.7	
8.8	Payback time (after income tax)	years	27.6	
8.9	FIRR of project capital fund	%	2.09	
8.10	FIRR of enterprise capital fund	%	8.00	
9	Loan period	a	25	
10	Financing situation			
10.1	Capital fund	10^8 yuan	13.21	Excluding working capital
	Government investment	10^8 yuan	9.99	Excluding working capital
	Enterprise capital fund	10^8 yuan	3.22	Excluding working capital
10.2	Proportion of capital to fixed assets investment	%	64.94	
10.3	Proportion of capital to total investment	%	62.23	
10.4	Loan capital fund	10^8 yuan	7.13	Excluding working capital

10.9 Financial Evaluation

Continued

Serial number	Item	Unit	Indicators	Note
10.5	Proportion of loan principal to fixed assets investment	%	35.06	
10.6	Loan interest	10^8 yuan	0.89	Excluding working capital
10.7	Total loan principal and interest	10^8 yuan	8.02	Excluding working capital
10.8	Proportion of total loan principal and interest to total investment	%	37.85	

1. Financial viability analysis

In the early stage of project operation, especially in the early five years, the water demand in the receiving area does not reach the designed value, while the pressure of repayment is great. Therefore, short-term loans from banks are necessary. After the 6th year, the net cash flow is greater than 0, which can ensure the normal operation of the project. Therefore, the project has financial viability. See Table 10.21 for the cash flow of financial plan.

2. Solvency analysis

The loan of this project shall be repaid in the form of fixed payment mortgage. The principal funds for repayment of loans include undistributed profits and depreciation costs. The loan repayment period is 25 years (including the construction period), and the repayment plan is provided in Table 10.22. The interest reserve rate is greater than 1, and it is greater than 2 after the 15th year. Due to small water consumption during the 5th-7th year, the project can not gain much benefit and the debt service reserve rate is less than 1. Therefore, a short-term loan is needed to repay the long-term principal. With the increase of water supply in the 8th-10th year, the short-term loans can be repaid, and the interest reserve rate after the 11th year is greater than 1.3.

The asset-liability ratio is the highest, i.e., 38.85%, in the 2nd year of project operation and then decreases year by year. In the 26th year, the long-term loan principal and interest are all paid off, and the asset-liability ratio is less than 1%. Therefore, the project has certain debt paying ability. The financial balancing sheet is provided in Table 10.23.

3. Profitability analysis

The financial internal rate of return (FIRR) of all investment, capital fund, and enterprise capital fund in the project are 2.66% (after income tax), 2.09%, and 8%, respectively. The investment payoff period is 27.8 years (after income tax). The total return on investment is 2.82%. The net profit rate of the capital is 2.55%.

See Table 10.24 for cash flow of project investment, Table 10.25 for financial cash flow of project capital, and Table 10.26 for cash flow of enterprise capital.

10 Case Study of Economic Evaluation of Water Conservancy Projects

Table 10.21 Cash flow statement of financial plan

unit: 10^4 yuan

Serial number	Item	Total	1	2	3	4	5	6	7	8	9	10	11	12	13	14	15	16	17	18	19	20	21
1	Net cash from operating activities	382303	0	0	0	0	5110	5755	6401	7047	7693	8339	8945	9388	9829	10268	10704	10653	10599	10542	10482	10418	10351
1.1	Cash income	632617	0	0	0	0	9222	9987	10751	11516	12280	13044	13809	14573	15338	16102	16866	16866	16866	16866	16866	16866	16866
1.1.1	Sales income	632617	0	0	0	0	9222	9987	10751	11516	12280	13044	13809	14573	15338	16102	16866	16866	16866	16866	16866	16866	16866
1.1.2	Substituted money on VAT	0																						
1.1.3	Subsidy income	0	0	0	0	0	0	0	0	0	0	0	0	0	0	0	0	0	0	0	0	0	0
1.1.4	Other income	0	0	0	0	0	0	0	0	0	0	0	0	0	0	0	0	0	0	0	0	0	0
1.2	Cash outflow	250314	0	0	0	0	4113	4231	4350	4468	4587	4705	4864	5185	5508	5834	6162	6213	6267	6324	6385	6448	6515
1.2.1	Annual operation cost	178830	0	0	0	0	3811	3888	3964	4041	4117	4194	4270	4347	4423	4499	4576	4576	4576	4576	4576	4576	4576
1.2.2	Input value added tax	0																						
1.2.3	Sales tax and surcharges	26564	0	0	0	0	301	344	386	428	470	512	554	596	638	680	722	722	722	722	722	722	722
1.2.4	Value added tax	0																						
1.2.5	Income tax	44920	0	0	0	0	0	0	0	0	0	0	40	243	448	655	865	916	970	1027	1087	1150	1218
1.2.6	Other outflows	0	0	0	0	0	0	0	0	0	0	0	0	0	0	0	0	0	0	0	0	0	0
2	Net cash flow from investment activities	−215790	−48736	−57212	−52920	−44489	−445	0	0	0	0	0	0	0	0	0	0	0	0	0	0	0	0
2.1	Cash inflow	0																						
2.2	Cash outflow	215790	48736	57212	52920	44489	445	0	0	0	0	0	0	0	0	0	0	0	0	0	0	0	0
2.2.1	Fixed-asset investment	203357	48736	57212	52920	44489	0	0	0	0	0	0	0	0	0	0	0	0	0	0	0	0	0
2.2.2	Renovation investment	11988	0	0	0	0	0	0	0	0	0	0	0	0	0	0	0	0	0	0	0	0	0
2.2.3	Working capital	445					445																	
2.2.4	Other outflows	0																						
3	Net cash flow from financing activities	75377	48736	57212	52920	44489	−4665	−5755	−6401	−7047	−7693	−8337	−6630	−6630	−6630	−6630	−6630	−6630	−6630	−6630	−6630	−6630	−6630

10.9 Financial Evaluation

Continued

Serial number	Item	Total	1	2	3	4	5	6	7	8	9	10	11	12	13	14	15	16	17	18	19	20	21
3.1	Cash inflow	235660	49247	58845	55742	48391	1966	2473	2829	2556	1623	0	0	0	0	0	0	0	0	0	0	0	0
3.1.1	Project capital investment	144189	30646	36619	34688	30114	134	0	0	0	0	0	0	0	0	0	0	0	0	0	0	0	0
3.1.2	Project investment loan	80157	18601	22226	21054	18277	0	0	0	0	0	0	0	0	0	0	0	0	0	0	0	0	0
3.1.3	Short term loan	11002	0	0	0	0	1521	2473	2829	2556	1623	0	0	0	0	0	0	0	0	0	0	0	0
3.1.4	Bond	0																					
3.1.5	Working capital loan	312	0	0	0	0	312	0	0	0	0	0	0	0	0	0	0	0	0	0	0	0	0
3.1.6	Other inflow	0																					
3.2	Cash outflow	160283	511	1633	2822	3902	6630	8229	9230	9603	9317	8337	6630	6630	6630	6630	6630	6630	6630	6630	6630	6630	6630
3.2.1	Repayment of long-term loan principal	80157	0	0	0	0	2086	2203	2328	2460	2598	2745	2900	3064	3237	3420	3614	3818	4033	4261	4502	4756	5025
3.2.2	Repayment of short-term loan principal	11002	0	0	0	0	0	1521	2473	2829	2556	1623	0	0	0	0	0	0	0	0	0	0	0
3.2.3	Bond repayment	0																					
3.2.4	Repayment of working capital loan principal	312	0	0	0	0	0	0	0	0	0	0	0	0	0	0	0	0	0	0	0	0	0
3.2.5	Long-term loan interest expenditure		511	1633	2822	3902	4529	4411	4287	4155	4016	3869	3714	3550	3377	3194	3001	2797	2581	2353	2112	1858	1589
3.2.6	Short-term loan interest expenditure	561	0	0	0	0	0	78	126	144	130	83	0	0	0	0	0	0	0	0	0	0	0
3.2.7	Interest expense of working capital loan	635	0	0	0	0	16	16	16	16	16	16	16	16	16	16	16	16	16	16	16	16	16
3.2.8	Profitpayable	0	0	0	0	0	0	0	0	0	0	0	0	0	0	0	0	0	0	0	0	0	0
3.2.9	Other outflows	0																					
4	Net cash flows												2314	2758	3199	3638	4074	4023	3969	3912	3852	3788	3721
5	Accumulated surplus capital fund	241890	0	0	0	0	0	0	0	0	0	2	2317	5074	8273	11911	15984	20007	23976	27888	31739	35527	39248

10 Case Study of Economic Evaluation of Water Conservancy Projects

Continued

Serial number	Item	22	23	24	25	26	27	28	29	30	31	32	33	34	35	36	37	38	39	40	41	42	43	44
1	Net cash from operating activities	10280	10205	10126	10042	9954	9954	9954	9954	9954	9954	9954	9954	9954	9954	9954	9954	9954	9954	9954	9954	9954	9954	9954
1.1	Cash income	16866	16866	16866	16866	16866	16866	16866	16866	16866	16866	16866	16866	16866	16866	16866	16866	16866	16866	16866	16866	16866	16866	16866
1.1.1	Sales income	16866	16866	16866	16866	16866	16866	16866	16866	16866	16866	16866	16866	16866	16866	16866	16866	16866	16866	16866	16866	16866	16866	16866
1.1.2	Substituted money on VAT	0	0	0	0	0	0	0	0	0	0	0	0	0	0	0	0	0	0	0	0	0	0	0
1.1.3	Subsidy income	0	0	0	0	0	0	0	0	0	0	0	0	0	0	0	0	0	0	0	0	0	0	0
1.1.4	Other income	0	0	0	0	0	0	0	0	0	0	0	0	0	0	0	0	0	0	0	0	0	0	0
1.2	Cash outflow	6586	6661	6740	6824	6913	6913	6913	6913	6913	6913	6913	6913	6913	6913	6913	6913	6913	6913	6913	6913	6913	6913	6913
1.2.1	Annual operation cost	4576	4576	4576	4576	4576	4576	4576	4576	4576	4576	4576	4576	4576	4576	4576	4576	4576	4576	4576	4576	4576	4576	4576
1.2.2	Input value added tax	0	0	0	0	0	0	0	0	0	0	0	0	0	0	0	0	0	0	0	0	0	0	0
1.2.3	Sales tax and surcharges	722	722	722	722	722	722	722	722	722	722	722	722	722	722	722	722	722	722	722	722	722	722	722
1.2.4	Value added tax	0	0	0	0	0	0	0	0	0	0	0	0	0	0	0	0	0	0	0	0	0	0	0
1.2.5	Income tax	1289	1363	1443	1526	1615	1615	1615	1615	1615	1615	1615	1615	1615	1615	1615	1615	1615	1615	1615	1615	1615	1615	1615
1.2.6	Other outflows	0	0	0	0	0	0	0	0	0	0	0	0	0	0	0	0	0	0	0	0	0	0	0
2	Net cash flow from investment activities	−6630	−6630	−6630	−6630	−16	−16	−5994	−5994	−16	−16	−16	−16	−16	−16	−16	−16	−16	−16	−16	−16	−16	−16	−16
2.1	Cash inflow																							
2.2	Cash outflow	0	0	0	0	0	0	5994	5994	0	0	0	0	0	0	0	0	0	0	0	0	0	0	0
2.2.1	Fixed-asset investment	0	0	0	0	0	0	0	0	0	0	0	0	0	0	0	0	0	0	0	0	0	0	0
2.2.2	Renovation investment	0	0	0	0	0	0	5994	5994	0	0	0	0	0	0	0	0	0	0	0	0	0	0	0
2.2.3	Working capital	0	0	0	0	0	0	0	0	0	0	0	0	0	0	0	0	0	0	0	0	0	0	0
2.2.4	Other outflows																							
3	Net cash flow from financing activities	−6630	−6630	−6630	−6630	−16	−16	5978	5978	−16	−16	−16	−16	−16	−16	−16	−16	−16	−16	−16	−16	−16	−16	−327

10.9 Financial Evaluation

Continued

Serial number	Item	22	23	24	25	26	27	28	29	30	31	32	33	34	35	36	37	38	39	40	41	42	43	44
3.1	Cash inflow	0	0	0	0	0	0	5994	5994	0	0	0	0	0	0	0	0	0	0	0	0	0	0	0
3.1.1	Project capital investment	0	0	0	0	0	0	5994	5994	0	0	0	0	0	0	0	0	0	0	0	0	0	0	0
3.1.2	Project investment loan	0	0	0	0	0	0	0	0	0	0	0	0	0	0	0	0	0	0	0	0	0	0	0
3.1.3	Short term loan	0	0	0	0	0	0	0	0	0	0	0	0	0	0	0	0	0	0	0	0	0	0	0
3.1.4	Bond																							
3.1.5	Working capital loan	0	0	0	0	0	0	0	0	0	0	0	0	0	0	0	0	0	0	0	0	0	0	0
3.1.6	Other inflow																							
3.2	Cash outflow	6630	6630	6630	6630	16	16	16	16	16	16	16	16	16	16	16	16	16	16	16	16	16	16	327
3.2.1	Repayment of long-term loan principal	5309	5609	5926	6261	0	0	0	0	0	0	0	0	0	0	0	0	0	0	0	0	0	0	0
3.2.2	Repayment of short-term loan principal	0	0	0	0	0	0	0	0	0	0	0	0	0	0	0	0	0	0	0	0	0	0	0
3.2.3	Bond repayment																							
3.2.4	Repayment of working capital loan principal	0	0	0	0	0	0	0	0	0	0	0	0	0	0	0	0	0	0	0	0	0	0	312
3.2.5	Long-term loan interest expenditure	1305	1005	689	354	0	0	0	0	0	0	0	0	0	0	0	0	0	0	0	0	0	0	0
3.2.6	Short-term loan interest expenditure	0	0	0	0	0	0	0	0	0	0	0	0	0	0	0	0	0	0	0	0	0	0	0
3.2.7	Interest expense of working capital loan	16	16	16	16	16	16	16	16	16	16	16	16	16	16	16	16	16	16	16	16	16	16	16
3.2.8	Profit payable	0	0	0	0	0	0	0	0	0	0	0	0	0	0	0	0	0	0	0	0	0	0	0
3.2.9	Other outflows																							
4	Net cash flows	3650	3575	3496	3412	9938	9938	9938	9938	9938	9938	9938	9938	9938	9938	9938	9938	9938	9938	9938	9938	9938	9938	9626
5	Accumulated surplus capital fund	42898	46473	49968	53380	63318	73256	83194	93132	103070	113008	122946	132884	142822	152760	162698	172636	182574	192512	202450	212388	222326	232264	241890

247

Table 10.22　Schedule of loan repayment of principal and interest

unit: 10⁴ yuan

Serial number	Item	Total	1	2	3	4	5	6	7	8	9	10	11	12	13	14	15	16	17	18	19	20	21	22	23	24	25
1	Loan and repayment of principal and interest						Grace period	4		Loan term	25		Loan rate	5.65%													
1.1	Accumulated principal and interest at the beginning of the year		0	18601	40826	61880	80157	78072	75868	73540	71081	68482	65737	62837	59772	56535	53115	49501	45683	41650	37389	32887	28130	23105	17796	12187	6261
1.1.1	Principal		0	18090	38683	56914	71290	78072	75868	73540	71081	68482	65737	62837	59772	56535	53115	49501	45683	41650	37389	32887	28130	23105	17796	12187	6261
1.1.2	Interest		0	511	2144	4965	8868	0	0	0	0	0	0	0	0	0	0	0	0	0	0	0	0	0	0	0	0
1.2	Loan of this year		18090	20593	18232	14375	0	0	0	0	0	0	0	0	0	0	0	0	0	0	0	0	0	0	0	0	0
1.3	Accrued interest of the year		511	1633	2822	3902	4529	4411	4287	4155	4016	3869	3714	3550	3377	3194	3001	2797	2581	2353	2112	1858	1589	1305	1005	689	354
1.4	Repayment of principal and interest	138905	0	0	0	0	6615	6615	6615	6615	6615	6615	6615	6615	6615	6615	6615	6615	6615	6615	6615	6615	6615	6615	6615	6615	6615
1.4.1	Principal repayment of this year		0	0	0	0	2086	2203	2328	2460	2598	2745	2900	3064	3237	3420	3614	3818	4033	4261	4502	4756	5025	5309	5609	5926	6261
1.4.2	Interest payment of this year		0	0	0	0	4529	4411	4287	4155	4016	3869	3714	3550	3377	3194	3001	2797	2581	2353	2112	1858	1589	1305	1005	689	354
2	Source of repayment funds		0	0	0	0	5094	5662	6259	6887	7547	8240	8919	11604	14668	18110	21924	25672	29350	32954	36479	39922	43277	46541	49706	52769	55723
2.1	Undistributed profit		0	0	0	0	0	0	0	0	0	0	109	656	1209	1768	2334	2472	2618	2772	2934	3106	3287	3479	3681	3895	4121
2.2	Depreciation cost		0	0	0	0	565	1251	1973	2732	3531	4371	5093	5093	5093	5093	5093	5093	5093	5093	5093	5093	5093	5093	5093	5093	5093
2.3	Amortization fee		0	0	0	0	0	0	0	0	0	0	0	0	0	0	0	0	0	0	0	0	0	0	0	0	0

10.9 Financial Evaluation

Continued

Serial number	Item	Total	1	2	3	4	5	6	7	8	9	10	11	12	13	14	15	16	17	18	19	20	21	22	23	24	25
2.4	Other capital fund		0	0	0	0	0	0	0	0	0	0	0	0	0	0	0	0	0	0	0	0	0	0	0	0	0
2.5	Long-term loan interest expense included in cost		0	0	0	0	4529	4411	4287	4155	4016	3869	3714	3550	3377	3194	3001	2797	2581	2353	2112	1858	1589	1305	1005	689	354
2.6	Accumulated funds available for repayment in the previous year		0	0	0	0	0	0	0	0	0	0	2	2305	4989	8054	11495	15309	19057	22735	26339	29865	33307	36663	39926	43092	46155
3	Short term loan																										
3.1	Total principal and interest of initial short-term loan	11002	0	0	0	0	0	1521	2473	2829	2556	1623	0	0	0	0	0	0	0	0	0	0	0	0	0	0	0
3.2	New short-term loans	2829	0	0	0	0	1521	953	355	0	0	0	0	0	0	0	0	0	0	0	0	0	0	0	0	0	0
3.3	Short term loan interest	561	0	0	0	0	0	78	126	144	130	83	0	0	0	0	0	0	0	0	0	0	0	0	0	0	0
3.4	Repayment of short-term loans	2829	0	0	0	0	0	0	0	273	932	1623	0	0	0	0	0	0	0	0	0	0	0	0	0	0	0
3.5	Payment of short-term loan interest	561	0	0	0	0	0	78	126	144	130	83	0	0	0	0	0	0	0	0	0	0	0	0	0	0	0
	Calculation indicators																											
	Interest reserve rate						1.00	1.00	1.00	1.00	1.00	1.00	1.04	1.27	1.53	1.82	2.15	2.30	2.49	2.73	3.04	3.46	4.03	4.90	6.34	9.19	17.52
	Debt service coverage ratio						0.77	0.70	0.69	0.73	0.83	1.00	1.35	1.42	1.48	1.55	1.61	1.61	1.60	1.59	1.58	1.57	1.56	1.55	1.54	1.53	1.51

Table 10.23　Balance sheet

unit: 10^4 yuan

Serial number	Item	Total	1	2	3	4	5	6	7	8	9	10	11	12	13	14	15	16	17	18	19	20	21
1	Capital fund		49247	108092	163833	212225	207576	202483	197390	192296	187203	182112	179333	176997	175102	173647	172627	171556	170432	169250	168008	166703	165330
1.1	Working capital	3334590	0	0	0	0	445	445	445	445	445	447	2762	5519	8718	12356	16429	20452	24421	28333	32184	35972	39693
1.1.1	Monetary capital fund	17800	0	0	0	0	445	445	445	445	445	445	445	445	445	445	445	445	445	445	445	445	445
1.1.2	Account payable	0																					
1.1.3	Prepaid account	0																					
1.1.4	Stock	0																					
1.1.5	Others	3316790	0	0	0	0	0	0	0	0	0	2	2317	5074	8273	11911	15984	20007	23976	27888	31739	35527	39248
1.2	Project under construction	533397	49247	108092	163833	212225	0	0	0	0	0	0	0	0	0	0	0	0	0	0	0	0	0
1.3	Net value of fixed assets	4312407					207131	202038	196945	191851	186758	181664	176571	171478	166384	161291	156197	151104	146011	140917	135824	130730	125637
1.4	Net intangible and other assets	0					0	0	0	0	0	0	0	0	0	0	0	0	0	0	0	0	0
2	Liabilities and owners' equity	8180394	49247	108092	163833	212225	207576	202483	197390	192296	187203	182112	179333	176997	175102	173647	172627	171556	170432	169250	168008	166703	165330
2.1	Total amount of working capital	11002	0	0	0	0	1521	2473	2829	2556	1623	0	0	0	0	0	0	0	0	0	0	0	0
2.1.1	Short term loan	11002	0	0	0	0	1521	2473	2829	2556	1623	0	0	0	0	0	0	0	0	0	0	0	0
2.1.2	Account payable	0																						
2.1.3	Prepaid account	0																						
2.1.4	Others	0																						
2.2	Project investment loan	1161090	18601	40826	61880	80157	78072	75868	73540	71081	68482	65737	62837	59772	56535	53115	49501	45683	41650	37389	32887	28130	23105
2.3	Short-term bank loan	12149	0	0	0	0	312	312	312	312	312	312	312	312	312	312	312	312	312	312	312	312	312
2.4	Subtotal of liabilities	1184241	18601	40826	61880	80157	79904	78653	76680	73948	70417	66048	63148	60084	56846	53426	49813	45995	41961	37700	33198	28442	23416
2.5	Owner's equity	6996153	30646	67265	101954	132067	127672	123830	120709	118348	116786	116063	116185	116913	118256	120221	122814	125561	128470	131550	134810	138261	141914

10.9　Financial Evaluation

Serial number	Item	Total	1	2	3	4	5	6	7	8	9	10	11	12	13	14	15	16	17	18	19	20	21
2.5.1	Capital fund	5619972	30646	67265	101954	132067	132201	132201	132201	132201	132201	132201	132201	132201	132201	132201	132201	132201	132201	132201	132201	132201	132201
2.5.2	Capital reserve	0																					
2.5.3	Accumulated surplus reserve fund		0	0	0	0	0	0	0	0	0	0	12	85	219	416	675	950	1241	1549	1875	2220	2585
2.5.4	Accumulated undistributed profits		0	0	0	0	−4529	−8371	−11492	−13853	−15415	−16138	−16028	−15373	−14164	−12396	−10062	−7589	−4972	−2200	734	3840	7128
	Asset liability ratio/%		37.77	37.77	37.77	37.77	38.49	38.84	38.85	38.46	37.62	36.27	35.21	33.95	32.46	30.77	28.86	26.81	24.62	22.27	19.76	17.06	14.16

Serial number	Item	Total	22	23	24	25	26	27	28	29	30	31	32	33	34	35	36	37	38	39	40	41	42	43	44
1	Capital fund	163886	162368	160770	159089	163933	168778	173622	178467	183311	188156	193001	197845	202690	207534	212379	217224	222068	226913	231757	236602	241446	246291	250824
1.1	Working capital	43343	46918	50413	53825	63763	73701	83639	93577	103515	113453	123391	133329	143267	153205	163143	173081	183019	192957	202895	212833	222771	232709	242335
1.1.1	Monetary capital fund	445	445	445	445	445	445	445	445	445	445	445	445	445	445	445	445	445	445	445	445	445	445	445
1.1.2	Account payable																								
1.1.3	Prepaid account																								
1.1.4	Stock																								
1.1.5	Others	42898	46473	49968	53380	63318	73256	83194	93132	103070	113008	122946	132884	142822	152760	162698	172636	182574	192512	202450	212388	222326	232264	241890
1.2	Project under construction	120544	115450	110357	105263	100170	95077	89983	84890	79797	74703	69610	64516	59423	54330	49236	44143	39049	33956	28863	23769	18676	13582	8489
1.3	Net value of fixed assets	0	0	0	0	0	0	0	0	0	0	0	0	0	0	0	0	0	0	0	0	0	0	0
1.4	Net intangible and other assets	0	0	0	0	0	0	0	0	0	0	0	0	0	0	0	0	0	0	0	0	0	0	0

Continued

Continued

Serial number	Item	22	23	24	25	26	27	28	29	30	31	32	33	34	35	36	37	38	39	40	41	42	43	44
2	Liabilities and owners' equity	163886	162368	160770	159089	163933	168778	173622	178467	183311	188156	193001	197845	202690	207534	212379	217224	222068	226913	231757	236602	241446	246291	250824
2.1	Total amount of working capital	0	0	0	0	0	0	0	0	0	0	0	0	0	0	0	0	0	0	0	0	0	0	0
2.1.1	Short term loan	0	0	0	0	0	0	0	0	0	0	0	0	0	0	0	0	0	0	0	0	0	0	0
2.1.2	Account payable																							
2.1.3	Prepaid account																							
2.1.4	Others																							
2.2	Project investment loan	17796	12187	6261	0	0	0	0	0	0	0	0	0	0	0	0	0	0	0	0	0	0	0	0
2.3	Short-term bank loan	312	312	312	312	312	312	312	312	312	312	312	312	312	312	312	312	312	312	312	312	312	312	0
2.4	Subtotal of liabilities	18107	12498	6572	312	312	312	312	312	312	312	312	312	312	312	312	312	312	312	312	312	312	312	0
2.5	Owner's equity	145779	149870	154198	158777	163622	168466	173311	178155	183000	187845	192689	197534	202378	207223	212067	216912	221757	226601	231446	236290	241135	245979	250824
2.5.1	Capital fund	132201	132201	132201	132201	132201	132201	132201	132201	132201	132201	132201	132201	132201	132201	132201	132201	132201	132201	132201	132201	132201	132201	132201
2.5.2	Capital reserve																							
2.5.3	Accumulated surplus reserve fund	2972	3381	3813	4271	4756	5240	5725	6209	6694	7178	7663	8147	8631	9116	9600	10085	10569	11054	11538	12023	12507	12992	13476
2.5.4	Accumulated undistributed profits	10607	14288	18183	22305	26665	31025	35385	39745	44105	48465	52826	57186	61546	65906	70266	74626	78986	83346	87707	92067	96427	100787	105147
	Asset liability ratio/%	11.05	7.70	4.09	0.20	0.19	0.18	0.18	0.17	0.17	0.17	0.16	0.16	0.15	0.15	0.15	0.14	0.14	0.14	0.13	0.13	0.13	0.13	0

10.9 Financial Evaluation

Table 10.24　Cash flow statement of project all investment

unit: 10^4 yuan

Serial number	Item	Total	1	2	3	4	5	6	7	8	9	10	11	12	13	14	15	16	17	18	19	20	21
1	Cash inflow	641551	0	0	0	0	9222	9987	10751	11516	12280	13044	13809	14573	15338	16102	16866	16866	16866	16866	16866	16866	16866
1.1	Sales income	632617	0	0	0	0	9222	9987	10751	11516	12280	13044	13809	14573	15338	16102	16866	16866	16866	16866	16866	16866	16866
1.2	Service income	0																					
1.3	Subsidy income	0	0	0	0	0	0	0	0	0	0	0	0	0	0	0	0	0	0	0	0	0	0
1.4	Recovery of residual value of fixed assets	8489																						5298
1.5	Recovery of working capital	445					0	0	0	0	0	0	0	0	0	0	0	0	0	0	0	0	0
2	Cash outflow	421184	48736	57212	52920	44489	4558	4231	4350	4468	4587	4705	4824	4942	5061	5179	5298	5298	5298	5298	5298	5298	5298
2.1	Investment in fixed assets	203357	48736	57212	52920	44489	0	0	0	0	0	0	0	0	0	0	0	0	0	0	0	0	0
2.2	Working capital	445	0	0	0	0	445	0	0	0	0	0	0	0	0	0	0	0	0	0	0	0	0
2.3	Annual operation cost	178830	0	0	0	0	3811	3888	3964	4041	4117	4194	4270	4347	4423	4499	4576	4576	4576	4576	4576	4576	4576
2.4	Sales tax and surcharges	26564	0	0	0	0	301	344	386	428	470	512	554	596	638	680	722	722	722	722	722	722	722
2.5	Renovation investment	11988	0	0	0	0	0	0	0	0	0	0	0	0	0	0	0	0	0	0	0	0	0
3	Net cash flow before income tax	220367	−48736	−57212	−52920	−44489	4665	5755	6401	7047	7693	8339	8985	9631	10277	10923	11569	11569	11569	11569	11569	11569	11569
4	Accumulated net cash flow before income tax		−48736	−105948	−158868	−203357	−198692	−192937	−186536	−179488	−171795	−163456	−154471	−144840	−134563	−123640	−112072	−100503	−88934	−77365	−65797	−54228	−42659
5	Income tax adjustment	59906	0	0	0	0	1136	1126	1107	1079	1041	992	973	1134	1296	1457	1619	1619	1619	1619	1619	1619	1619
6	Net cash flow after income tax	160461	−48736	−57212	−52920	−44489	3528	4629	5294	5969	6653	7347	8012	8497	8981	9465	9950	9950	9950	9950	9950	9950	9950

10 Case Study of Economic Evaluation of Water Conservancy Projects

Continued

Serial number	Item	22	23	24	25	26	27	28	29	30	31	32	33	34	35	36	37	38	39	40	41	42	43	44
1	Cash inflow	16866	16866	16866	16866	16866	16866	16866	16866	16866	16866	16866	16866	16866	16866	16866	16866	16866	16866	16866	16866	16866	16866	25800
1.1	Sales income	16866	16866	16866	16866	16866	16866	16866	16866	16866	16866	16866	16866	16866	16866	16866	16866	16866	16866	16866	16866	16866	16866	16866
1.2	Service income																							
1.3	Subsidy income	0	0	0	0	0	0	0	0	0	0	0	0	0	0	0	0	0	0	0	0	0	0	0
1.4	Recovery of total value of fixed assets	0	0	0	0	0	0	0	0	0	0	0	0	0	0	0	0	0	0	0	0	0	0	8489
1.5	Recovery of working capital	0	0	0	0	0	0	0	0	0	0	0	0	0	0	0	0	0	0	0	0	0	0	445
2	Cash outflow	5298	5298	5298	5298	5298	5298	11292	11292	5298	5298	5298	5298	5298	5298	5298	5298	5298	5298	5298	5298	5298	5298	5298
2.1	Fixed asset investment	0	0	0	0	0	0	0	0	0	0	0	0	0	0	0	0	0	0	0	0	0	0	0
2.2	Working capital	0	0	0	0	0	0	0	0	0	0	0	0	0	0	0	0	0	0	0	0	0	0	0
2.3	Annual operation cost	4576	4576	4576	4576	4576	4576	4576	4576	4576	4576	4576	4576	4576	4576	4576	4576	4576	4576	4576	4576	4576	4576	4576
2.4	Sales tax and surcharges	722	722	722	722	722	722	722	722	722	722	722	722	722	722	722	722	722	722	722	722	722	722	722
2.5	Renovation investment	0	0	0	0	0	0	5994	5994	0	0	0	0	0	0	0	0	0	0	0	0	0	0	0
3	Net cash inflow before income tax	11569	11569	11569	11569	11569	11569	5575	5575	11569	11569	11569	11569	11569	11569	11569	11569	11569	11569	11569	11569	11569	11569	20503
4	Net cash flow before accumulated income tax	−31090	−19522	−7953	3616	15184	26753	32328	37903	49471	61040	72609	84177	95746	107315	118884	130452	142021	153590	165159	176727	188296	199865	220367
5	Income tax adjustment	1619	1619	1619	1619	1619	1619	1619	1619	1619	1619	1619	1619	1619	1619	1619	1619	1619	1619	1619	1619	1619	1619	1619
6	Net cash flow after income tax	9950	9950	9950	9950	9950	9950	3956	3956	9950	9950	9950	9950	9950	9950	9950	9950	9950	9950	9950	9950	9950	9950	18884

Evaluation indicators: before income tax, after income tax.

FIRR of all investment: 3.45%, 2.66%.

FNPV of all investment: −883.51 million yuan, −1000.88 million yuan.

Payback period of total investment: 24.7 years, 27.6 years.

10.9 Financial Evaluation

Table 10.25　Capital cash flow statement

unit: 10^4 yuan

Serial number	Item	Total	1	2	3	4	5	6	7	8	9	10	11	12	13	14	15	16	17	18	19	20	21
1	Cash inflow	652554	0	0	0	0	10743	12460	13580	14072	13904	13044	13809	14573	15338	16102	16866	16866	16866	16866	16866	16866	16866
1.1	Sales income	632617	0	0	0	0	9222	9987	10751	11516	12280	13044	13809	14573	15338	16102	16866	16866	16866	16866	16866	16866	16866
1.2	Service income	0																					
1.3	Subsidy income	0	0	0	0	0	0	0	0	0	0	0	0	0	0	0	0	0	0	0	0	0	0
1.4	Recovery of residual value of fixed assets	8489	0	0	0	0	0	0	0	0	0	0	0	0	0	0	0	0	0	0	0	0	0
1.5	Recovery of working capital	445	0	0	0	0	0	0	0	0	0	0	0	0	0	0	0	0	0	0	0	0	0
1.6	Income from short-term loans during operating period	11002	0	0	0	0	1521	2473	2829	2556	1623	0	0	0	0	0	0	0	0	0	0	0	0
2	Cash outflow	545918	30646	36619	34688	30114	10877	12460	13580	14072	13904	13042	11495	11816	12139	12465	12793	12844	12898	12955	13015	13079	13146
2.1	Project capital fund	132201	30646	36619	34688	30114	134	0	0	0	0	0	0	0	0	0	0	0	0	0	0	0	0
2.2	Repayment of loan principal	91471	0	0	0	0	2086	3724	4801	5288	5154	4369	2900	3064	3237	3420	3614	3818	4033	4261	4502	4756	5025
	Including: long-term loan	80157	0	0	0	0	2086	2203	2328	2460	2598	2745	2900	3064	3237	3420	3614	3818	4033	4261	4502	4756	5025
	Working capital loan	312	0	0	0	0	0	0	0	0	0	0	0	0	0	0	0	0	0	0	0	0	0
		11002	0	0	0	0	0	1521	2473	2829	2556	1623	0	0	0	0	0	0	0	0	0	0	0
2.3	Loan	59944	0	0	0	0	4545	4504	4429	4315	4162	3968	3730	3566	3393	3210	3017	2813	2597	2369	2128	1874	1605
	Including: long-term loan	58748	0	0	0	0	4529	4411	4287	4155	4016	3869	3714	3550	3377	3194	3001	2797	2581	2353	2112	1858	1589
	Working capital loan	635	0	0	0	0	16	16	16	16	16	16	16	16	16	16	16	16	16	16	16	16	16
	Short term loan	561	0	0	0	0	0	78	126	144	130	83	0	0	0	0	0	0	0	0	0	0	0
2.4	Annual operation cost	178830	0	0	0	0	3811	3888	3964	4041	4117	4194	4270	4347	4423	4499	4576	4576	4576	4576	4576	4576	4576
2.5	Sales income and surcharges	26564	0	0	0	0	301	344	386	428	470	512	554	596	638	680	722	722	722	722	722	722	722
2.6	Income tax	44920	0	0	0	0	0	0	0	0	0	0	40	243	448	655	865	916	970	1027	1087	1150	1218
2.7	Repayment of loan principal	11988	0	0	0	0	0	0	0	0	0	0	0	0	0	0	0	0	0	0	0	0	0
3	Net cash flow	106635	−30646	−36619	−34688	−30114	−134	0	0	0	0	2	2314	2758	3199	3638	4074	4023	3969	3912	3852	3788	3721

10 Case Study of Economic Evaluation of Water Conservancy Projects

Continued

Serial number	Item	22	23	24	25	26	27	28	29	30	31	32	33	34	35	36	37	38	39	40	41	42	43	44
1	Cash inflow	16866	16866	16866	16866	16866	16866	16866	16866	16866	16866	16866	16866	16866	16866	16866	16866	16866	16866	16866	16866	16866	16866	25800
1.1	Sales income	16866	16866	16866	16866	16866	16866	16866	16866	16866	16866	16866	16866	16866	16866	16866	16866	16866	16866	16866	16866	16866	16866	16866
1.2	Service income																							
1.3	Subsidy income	0	0	0	0	0	0	0	0	0	0	0	0	0	0	0	0	0	0	0	0	0	0	0
1.4	Recovery of residual value of fixed assets	0	0	0	0	0	0	0	0	0	0	0	0	0	0	0	0	0	0	0	0	0	0	8489
1.5	Recovery of working capital	0	0	0	0	0	0	0	0	0	0	0	0	0	0	0	0	0	0	0	0	0	0	445
1.6	Income from short-term loans during operating period	0	0	0	0	0	0	0	0	0	0	0	0	0	0	0	0	0	0	0	0	0	0	0
2	Cash outflow	13217	13292	13371	13455	6929	6929	12923	12923	6929	6929	6929	6929	6929	6929	6929	6929	6929	6929	6929	6929	6929	6929	7240
2.1	Project capital fund	0	0	0	0	0	0	0	0	0	0	0	0	0	0	0	0	0	0	0	0	0	0	0
2.2	Repayment of loan principal	5309	5609	5926	6261	0	0	0	0	0	0	0	0	0	0	0	0	0	0	0	0	0	0	312
	Including: long-term loan	5309	5609	5926	6261	0	0	0	0	0	0	0	0	0	0	0	0	0	0	0	0	0	0	0
	Working capital loan	0	0	0	0	0	0	0	0	0	0	0	0	0	0	0	0	0	0	0	0	0	0	312
2.3	Loan	1321	1021	704	370	0	0	0	0	0	0	0	0	0	0	0	0	0	0	0	0	0	0	0
	Including: long-term loan	1305	1005	689	354	0	0	0	0	0	0	0	0	0	0	0	0	0	0	0	0	0	0	0
	Working capital loan	16	16	16	16	16	16	16	16	16	16	16	16	16	16	16	16	16	16	16	16	16	16	16
	Short term loan	0	0	0	0	0	0	0	0	0	0	0	0	0	0	0	0	0	0	0	0	0	0	0
2.4	Annual operation cost	4576	4576	4576	4576	4576	4576	4576	4576	4576	4576	4576	4576	4576	4576	4576	4576	4576	4576	4576	4576	4576	4576	4576
2.5	Sales income and surcharges	722	722	722	722	722	722	722	722	722	722	722	722	722	722	722	722	722	722	722	722	722	722	722
2.6	Income tax	1289	1363	1443	1526	1615	1615	1615	1615	1615	1615	1615	1615	1615	1615	1615	1615	1615	1615	1615	1615	1615	1615	1615
2.7	Repayment of loan principal	0	0	0	0	0	0	5994	5994	0	0	0	0	0	0	0	0	0	0	0	0	0	0	0
3	Net cash flow	3650	3575	3496	3412	9938	9938	3944	3944	9938	9938	9938	9938	9938	9938	9938	9938	9938	9938	9938	9938	9938	9938	18560

FIRR of capital fund: 2.09%.

10.9 Financial Evaluation

Table 10.26　Cash flow statement of enterprise capital

unit: 10^4 yuan

Serial number	Item	Total	1	2	3	4	5	6	7	8	9	10	11	12	13	14	15	16	17	18	19	20	21
1	Cash inflow	652554	0	0	0	0	10743	12460	13580	14072	13904	13044	13809	14573	15338	16102	16866	16866	16866	16866	16866	16866	16866
1.1	Sales income	632617	0	0	0	0	9222	9987	10751	11516	12280	13044	13809	14573	15338	16102	16866	16866	16866	16866	16866	16866	16866
1.2	Service income	0																					
1.3	Subsidy income	0	0	0	0	0	0	0	0	0	0	0	0	0	0	0	0	0	0	0	0	0	0
1.4	Recovery of fixed assets	8489	0	0	0	0	0	0	0	0	0	0	0	0	0	0	0	0	0	0	0	0	0
1.5	Recovery of working capital	445	0	0	0	0	0	0	0	0	0	0	0	0	0	0	0	0	0	0	0	0	0
1.6	Income from short-term loans during operation period	11002	0	0	0	0	1521	2473	2829	2556	1623	0	0	0	0	0	0	0	0	0	0	0	0
2	Cash outflow	445908	7462	8917	8447	7333	10776	12460	13580	14072	13904	13042	11495	11816	12139	12465	12793	12844	12898	12955	13015	13079	13146
2.1	Enterprise capital	32191	7462	8917	8447	7333	33	0	0	0	0	0	0	0	0	0	0	0	0	0	0	0	0
2.2	Repayment of loan principal	91471	0	0	0	0	2086	3724	4801	5288	5154	4369	2900	3064	3237	3420	3614	3818	4033	4261	4502	4756	5025
	Including: Long-term loan	80157	0	0	0	0	2086	2203	2328	2460	2598	2745	2900	3064	3237	3420	3614	3818	4033	4261	4502	4756	5025
	Working capital	312	0	0	0	0	0	0	0	0	0	0	0	0	0	0	0	0	0	0	0	0	0
	Short term loan	11002	0	0	0	0	0	1521	2473	2829	2556	1623	0	0	0	0	0	0	0	0	0	0	0
2.3	Loan interest expenditure	59944	0	0	0	0	4545	4504	4429	4315	4162	3968	3730	3566	3393	3210	3017	2813	2597	2369	2128	1874	1605
	Including: Long-term loan	58748	0	0	0	0	4529	4411	4287	4155	4016	3869	3714	3550	3377	3194	3001	2797	2581	2353	2112	1858	1589
	Working capital loan	635	0	0	0	0	16	16	16	16	16	16	16	16	16	16	16	16	16	16	16	16	16
	Short-term loan	561	0	0	0	0	0	78	126	144	130	83	0	0	0	0	0	0	0	0	0	0	0
2.4	Operation cost	178830	0	0	0	0	3811	3888	3964	4041	4117	4194	4270	4347	4423	4499	4576	4576	4576	4576	4576	4576	4576
2.5	Sales tax and surcharges	26564	0	0	0	0	301	344	386	428	470	512	554	596	638	680	722	722	722	722	722	722	722
2.6	Income tax	44920	0	0	0	0	0	0	0	0	0	0	40	243	448	655	865	916	970	1027	1087	1150	1218
2.7	Renovation investment	11988	0	0	0	0	0	0	0	0	0	0	2314	2758	3199	3638	4074	4023	3969	3912	3852	3788	3721
3	Net cash flow	206645	−7462	−8917	−8447	−7333	−33	0	0	0	0	2	2314	2758	3199	3638	4074	4023	3969	3912	3852	3788	3721

Continued

Serial number	Item	22	23	24	25	26	27	28	29	30	31	32	33	34	35	36	37	38	39	40	41	42	43	44
1	Cash inflow	16866	16866	16866	16866	16866	16866	16866	16866	16866	16866	16866	16866	16866	16866	16866	16866	16866	16866	16866	16866	16866	16866	25800
1.1	Sales income	16866	16866	16866	16866	16866	16866	16866	16866	16866	16866	16866	16866	16866	16866	16866	16866	16866	16866	16866	16866	16866	16866	16866
1.2	Service income																							
1.3	Subsidy income	0	0	0	0	0	0	0	0	0	0	0	0	0	0	0	0	0	0	0	0	0	0	0
1.4	Recovery of fixed assets	0	0	0	0	0	0	0	0	0	0	0	0	0	0	0	0	0	0	0	0	0	0	8489
1.5	Recovery of working capital	0	0	0	0	0	0	0	0	0	0	0	0	0	0	0	0	0	0	0	0	0	0	445
1.6	Income from short-term loans during operation period	0	0	0	0	0	0	0	0	0	0	0	0	0	0	0	0	0	0	0	0	0	0	0
2	Cash outflow	13217	13292	13371	13455	6929	6929	12923	12923	6929	6929	6929	6929	6929	6929	6929	6929	6929	6929	6929	6929	6929	6929	7240
2.1	Enterprise capital	0	0	0	0	0	0	0	0	0	0	0	0	0	0	0	0	0	0	0	0	0	0	0
2.2	Repayment of loan principal	5309	5609	5926	6261	0	0	0	0	0	0	0	0	0	0	0	0	0	0	0	0	0	0	312
	Including: Long-term loan	5309	5609	5926	6261	0	0	0	0	0	0	0	0	0	0	0	0	0	0	0	0	0	0	0
	Working capital	0	0	0	0	0	0	0	0	0	0	0	0	0	0	0	0	0	0	0	0	0	0	312
	Short term loan	0	0	0	0	0	0	0	0	0	0	0	0	0	0	0	0	0	0	0	0	0	0	0
2.3	Loan interest expenditure	1321	1021	704	370	16	16	16	16	16	16	16	16	16	16	16	16	16	16	16	16	16	16	16
	Including: Long-term loan	1305	1005	689	354	0	0	0	0	0	0	0	0	0	0	0	0	0	0	0	0	0	0	0
	Working capital loan	16	16	16	16	16	16	16	16	16	16	16	16	16	16	16	16	16	16	16	16	16	16	16
	Short-term loan	0	0	0	0	0	0	0	0	0	0	0	0	0	0	0	0	0	0	0	0	0	0	0
2.4	Operation cost	4576	4576	4576	4576	4576	4576	4576	4576	4576	4576	4576	4576	4576	4576	4576	4576	4576	4576	4576	4576	4576	4576	4576
2.5	Sales tax and surcharges	722	722	722	722	722	722	722	722	722	722	722	722	722	722	722	722	722	722	722	722	722	722	722
2.6	Income tax	1289	1363	1443	1526	1615	1615	1615	1615	1615	1615	1615	1615	1615	1615	1615	1615	1615	1615	1615	1615	1615	1615	1615
2.7	Renovation investment	0	0	0	0	0	0	5994	5994	0	0	0	0	0	0	0	0	0	0	0	0	0	0	0
3	Net cash flow	3650	3575	3496	3412	9938	9938	3944	3944	9938	9938	9938	9938	9938	9938	9938	9938	9938	9938	9938	9938	9938	9938	18560

Calculation indicators: FIRR of Enterprise capital fund, 8.00%.

10.10 Conclusions

1. Financing plan

The recommended water supply price is 0.70 yuan/m³. The recommended on-grid price is 0.38 yuan/(kW·h). The recommended loan repayment period is 25 years. The financing plan of the project is as follows: the principal of the loan is 712.90 million yuan, the capital is 1320.67 million yuan, the interest rate during the construction period is 88.68 million yuan, and the total investment of the project is 2122.25 million yuan. The principal of the loan accounts for 35.06% of the fixed-asset investment, and the total loan accounts for 37.85% of the total investment.

Given the FIRR of enterprise capital as 8%, and the government capital paying no profits, the project can absorb enterprise investment of 321.58 million yuan. Correspondingly, the government investment is 999.09 million yuan.

2. Conclusion of economic analysis

The economic internal rate of return (EIRR) of the project is 14.24%, more than 8%. The economic net present value (ENPV, $is=8\%$) is 1577.89 million yuan, more than zero, so the project is economically reasonable.

3. Conclusion of financial evaluation

The financial internal rate of return of all investment in the project is 2.66% (after income tax). The financial internal rate of return of capital fund is 2.09%. And the financial internal rate of return of enterprise capital fund is 8%. The payback period of investment is 27.8 years (after income tax, including the construction period). The total return on investment is 2.82%, and the net profit margin of capital is 2.55%.

In the early stage of project operation, because the water demand in the receiving area of the project does not reach the design value and the repayment pressure in the early stage of operation is large, it is necessary to borrow short-term loans from the Bank 5 years before the project operation. After the 6th year of project operation, the net cash flow of project operation activities can ensure the normal operation of the project, and the project has the financial viability.

11

Foreign Case of Financial Evaluation of Water Conservancy and Hydropower Projects

The methods of financial evaluation and on-grid price estimation foreign of hydropower stations are much the same. Generally, the on-grid price of each year is calculated according to the requirements of project operation cost, repayment of principat and interest and owner's investment profit, and then the average on-grid price during the operating period is obtained, by considering the social discount rate. The on-grid price estination of Pakistan hydropower stations is highly representative. Therefore, the financial evaluation of foreign water conservancy and hydropower projects takes Pakistan hydropower station as an example. But it is worth mentioning that Pakistan hydropower projects have no interest tax, income tax and tumover tax. The following introdues the estimation method of on-grid price for hydropower stations and the financial evaluation of overseas financing projects.

11.1 Project Overview

QK Hydropower Station Project in Pakistan is in the North-west Frontier Province (currently Khyber Pakhtunkhwa). The project is a daily regulating hydropower station, which is developed in a mixed way. The project consists of river barrage, spillway, grit chamber, diversion tunnel, surge shaft, pressure pipe, powerhouse and tail race tunnel. The barrage is asphalt concrete face rockfill dam (maximum dam height=60 m, dam crest length=230 m). The storage capacity under high water level for normal operation of the reservoir is 12 million m³. The dead reservoir capacity is 4 million m³, and the regulating storage capacity is 3 million m³. The diversion tunnel is 18 km long (inner diameter= 7 m; rated discharge=120 m³/s). The installed capacity of the hydropower station is 800 MW, with four impulse units installed, with a maximum gross head of 780 m and long-term average annual energy output of 3450 GW · h.

The investment in the construction of this project is 1357.8 million US dollars (excluding the power transmission and transformation project). According to the construction schedule, the total construction period of the project is 5 years. The investment process and unit operating plan are shown in Table 11.1.

Table 11.1   **Investment Process and Unit Operating Plan**

Serial number	Item	Year						Total
		1	2	3	4	5	6 years later	
1	Construction investment /10^4 US dollars	13655	35828	35917	28552	21828		135780
2	Installed capacity at the end of year/MW					800	800	
3	Annual power generation/(GW·h)						3450	

11.2 Purpose and Basis of Financial Evaluation

11.2.1 Purpose of Financial Evaluation

The purpose of financial evaluation is to calculate the on-grid price and financial indicators and evaluate the financial feasibility, according to the requirements of relevant laws and regulations in hydropower industry in Pakistan, and the requirements of capital return rate.

11.2.2 Basis of Financial Evaluation

(1) various regulations, rules and guidelines issued by the Government of Pakistan for private investment projects and IPP projects, including:

1) *Pakistan Power Generation Projects Policy* (Year 2002).

2) Guide for determining IPPs on-grid price for independent power generators in Pakistan.

3) National Electric Power Regulatory Authority Mechanism for Determination of Tariff for Hydropower Projects.

4) Interim Electricity Procurement (procedures and standards) of the Pakistan Electric Power Regulatory Commission (version 2005).

5) Decision Letter for Reference Tariff of EPC Project.

(2) As a project invested by Chinese enterprise and financed by Bank of China, the evaluation basis also includes:

1) *Construction Project Economic Evaluation Methods and Parameter* (3rd Edition) (*fgtz* [2006] No. 1325).

2) *Specification on Economic Evaluation of Hydropower Project* (DL/T 5441-2010).

11.3 Tariff Structure of Pakistan

According to *Pakistan Power Generation Projects Policy (Year 2002)* and Guide for IPPs Price Determination, Pakistan's on-grid price consists of two parts, namely:

$$on\text{-}grid\ price = capacity\ price + energy\ price$$

According to "6.3.2 Capacity Price and Energy Price" of *Pakistan Power Generation Projects Policy (Year 2002)*: if the power output of the power station reaches the standard specified in the power purchase agreement, the capacity price will be paid. The energy price will be paid according to the actual power output of the power station. To ensure the interests of investors during the whole project life, the sum of energy price and capacity price will remain unchanged or increase over time, and the annual capacity price related to debt is the same as the loan repayment requirements of each year.

In Pakistan's electricity price system, capacity electricity price is composed of fixed operation and maintenance fee, insurance fee, debt service, capital return and capital return during construction, with the unit of Rs/(kW·month). Energy price is composed of variable operation and maintenance fee and water resources fee, with the unit of Rs/(kW·h), as shown in Table 11.2. The energy price only pays the operation cost of the new power generation of the power station. The capacity price bears the fixed operation and maintenance fee, insurance fee, repayment of principal and interest, return of capital and return of capital during the construction period of the project. Therefore, whether the power station can meet the power supply capacity required by the agreement is crucial to the financial revenue of the project.

Table 11.2　　Capacity price and energy price

Item	Capacity price	Energy price
Unit	Rs/(kW·month)	Rs/(kW·h)
Calculation basis	Net capacity	Actual on-grid energy
Components	Regular O&M fee Insurance premium Repayment of principal and interest Capital return Capital return during construction period Transfer and payment …	Variable O&M fee Water resources fee …

Hydropower is a kind of clean energy. According to Pakistan's electricity price policy, the capacity price of hydropower stations in Pakistan can account for more than 90% of the benchmark price.

11.4 Total Investment and Fund Raising of the Project

QK project is financed by loans from Chinese financial institutions. In addition to the payment of loan interest, it also needs to pay loan management fee, loan commitment fee and SINOSURE insurance premiums. The premiums include overseas loan insurance premiums and overseas investment insurance premiums.

1. construction investment

The construction investment of this project is 1357.8 million US dollars (excluding power transmission and transformation project).

2. loan interest

According to the requirements of the bank, the proportion of capital to the total investment should not be less than 25%. Herein, 25% of the total investment is adopted. The rest 75% of the total investment is borrowed from the bank. the loan interest rate is the sum of LIBOR and Margin. The average value of LIBOR is about 2.5% in the past 15 years. In this calculation, the Margin is 4.5%, and the loan interest rate is 7%. Herein, the annual interest rate is adopted.

According to the calculation, the interest rate on the loan during the construction period is 393.25 million US dollars, and the total principal and interest rate of the loan is 1313.29 million US dollars.

3. Overseas loan insurance premium

According to the requirements of the loan bank, when the Bank of China provides overseas investment loans to overseas projects in foreign countries, the project owner needs to purchase loan insurance from the China Export & Credit Insurance Cooperation. The premium rate of SINOSURE is 7%, and the calculation base is the sum of the loan principal and interest of the bank in the construction period and the loan interest rate during the operation period. The premium of the overseas loan is 130.89 million US dollars in the first year of operation.

4. Loan management fee

The loan management fee should be paid for the long-term US dollar loan applied to the bank. The loan management fee is based on the principal and interest of the loan during the construction period. The management fee rate is 1%, and the loan management fee is 13.13 million US dollars.

5. Loan commitment fee

For a long-term US dollar loan applied to a bank, the loan commitment fee shall be paid for the principal of the loan received. The loan commitment fee shall be based on the remaining loan amount. The loan commitment fee rate shall be 1%. Then the loan commitment fee shall be 19.34 million US dollars.

6. Overseas investment insurance premium

As this project is an overseas investment project, the project owner should buy insurance for the capital. The overseas loan insurance rate is 1%. Based on the accumulated investment of capital, the overseas investment insurance premium is 13.17 million US dollars.

7. Working capital

Assuming the installed capacity is 5 US dollars/kW, the working capital is 4 million US dollars. 25% is capital, and 75% uses loans. The loan interest rate of working capital is 7%. At the end of the calculation period, recover the working capital at one time.

8. Total investment

The interest of construction period is included, and the total investment of QK hydropower station is 1755.05 million US dollars (including working capital). See Table 11.3 for the total investment utilization plan and fund raising of the project.

Table 11.3   Project total investment use plan and fund-raising   unit: 10^4 US dollars yuan

Serial number	Item	Total	Construction period				
			1	2	3	4	5
1	Total investment	175505	29209	39040	41011	35481	30764
1.1	Construction investment	135780	13655	35828	35917	28552	21828
1.2	Bank loan interest	21672	222	2446	4481	6425	8097
1.3	Overseas loan insurance premium	13089	13089	0	0	0	0
1.4	Management fee	1313	1313	0	0	0	0
1.5	Commitment charges	1934	857	596	339	142	0
1.6	Overseas investment premium	1317	73	171	273	362	439
1.7	Working capital	400	0	0	0	0	400
2	Fund raising	175505	29209	39040	41011	35481	30764
2.1	Project capital	43876	7302	9760	10253	8870	7691
2.1.1	For construction investment	43776	7302	9760	10253	8870	7591
2.1.2	For working capital	100	0	0	0	0	100
2.2	Loan capital	131629	21907	29280	30758	26611	23073
2.2.1	For construction investment	92004	6353	26068	25664	19682	14237
2.2.2	For construction period interest	39325	15554	3212	5094	6929	8536
2.2.3	For working capital	300	0	0	0	0	300
2.3	Other capital	0	0	0	0	0	0

11.5 Total Cost Estimation

The total cost during the operation period of the project includes operation and management expenses (O&M), fixed assets insurance premiums, overseas investment insurance premiums, water resources utilization fees and interest expenses.

1. Operation and management expenses (O&M)

Operation and management expenses (O&M) include engineering maintenance expenses, annual salary of employees, material expenses and other expenses. Assume a rate of 1.5% of the construction investment, the O&M are $20.36 million. The O&M expenses include fixed and variable operation costs, accounting for 80% ($16.29 million) and 20% ($4.07 million), respectively. In the fixed operation costs, foreign and local currency accounts for 45% ($7.33 million) and 55% ($8.96 million), respectively. Comparatively, in the variable operating cost, the foreign and local currency accounts for 20% ($0.81 million) and 80% ($3.26 million).

According to historical statistics, Pakistan's inflation rate is 1.836% and the US dollar's inflation rate is 2.396%.

2. Property insurance premium

The property insurance premium is calculated as 1% of the construction investment, and the annual property insurance premium is 13.58 million US dollars.

3. Overseas investment insurance premium

The insurance premium for overseas investment is based on the net value of the project capital. According to the requirements of the insurance company, the rate is 1%.

4. Water use fee

According to *Pakistan Power Generation Projects Policy (Year 2002)*, the water resource management fee for power generation is charged at the rate of 0.15 rupees/(kW·h). This is equivalent to 0.00154/(kW·h) per US dollars given the exchange rate as 97.4 rupees per US dollar. The average annual water use fee is 5.5 million US dollars.

5. Interest expense

Generally, the term of overseas loans from Chinese financial institutions is not more than 20 years. The loan term is temporarily considered as 15 years, of which the grace period is 5 years and the payment period is 10 years. The interest rate of US dollar loan is 7%, in which LIBOR is 2.5% and Margin is 4.5%. The repayment method adopts the equal principal and interest method.

11.6 Calculation of Generation Income

Generation income = on-grid energy × on-grid price

The on-grid energy is the valid energy subtracting the auxiliary power consumption, and the transmission and transformation losses. Assume that the utilization coefficient of the valid energy is 0.95, and the ratio of the auxiliary power consumption is 1%. The on-grid power metering point is the input terminal of the transformer substation. Therefore, the transmission and transformation losses would not be considered.

The annual on-grid energy of the power station is 3245 GW · h.

11.7 Taxes

To stimulate the hydropower projects, according to *Pakistan Power Generation Projects Policy (Year 2002)*, the Pakistan government has formulated the following preferential policies for tax:

(1) Facilities, machinery and equipment used to produce electricity are exempt from sales tax.

(2) Income tax, business tax and withholding tax on imported products shall be exempted.

(3) The tariff of imported equipment is unified at 5%.

The tax not exempted by hydropower is the dividend tax of shareholders. Due to the uncertainty of the number of dividends, the government of Pakistan stipulates that 7.5% of capital gains shall be received in advance as withholding tax. However, this dividend withholding tax, as a "transfer payment" fee, is amortized into the capacity price and returned to the power supplier by the power purchaser in 12 months.

Above all, the annual sales tax and income tax of QK Hydropower Station is 0.

11.8 Capital Return Requirements

The on-grid price in Pakistan is determined based on the policy of "cost plus return". This allows the investors to obtain a reasonable rate of return on capital on the premise that investors can recover reasonable construction costs and operating costs. Pakistan has a maximum return on capital of 17% for private investment projects and IPP projects. The return on capital includes the return on capital during the construction period (ROEDC) and return on capital during operation (ROE).

1. Return on capital in construction period (ROEDC)

Return on capital in construction period is the return on capital invested in construction period that should have been paid in construction period. However, as there is no return on capital during construction period, therefore, the return on capital invested in each year is calculated to the end of construction period according to the requirement of return rate. Then, based on the return on capital, the return on capital is calculated for each year of operation. The return on capital during the construction period is calculated as follows:

$$ROEDC = -PMT(Rate, Nper, \Sigma ROEDC, Fv, Type)$$

Where:

PMT—the internal function of Excel, which is the equal installment repayment amount of loan under the fixed interest rate;

Rate—the return rate of capital;

Nper—the number of years of return on capital, which is taken as the project operation period;

$\Sigma ROEDC$—the sum of the return on capital during the construction period converted to the end of the construction period, which is calculated according to the following formula:

$$ROEDC_i = E \times [(1+Rate)^i - 1]$$

Where:

$ROEDC_i$—the amount of capital invested in year i to the end of the construction period after considering the rate of return;

E—the amount of capital invested in year i;

Rate—the return rate on capital.

2. Return on Capital During Operation (ROE)

The return on capital during the operation period is the return on capital invested by the owner during the operation period. It is calculated separately between the loan repayment period and the loan repayment period. The return on the capital during the loan repayment period is calculated according to the following formula:

$$ROE = E \cdot Rate$$

Where:

E—the amount of capital;

Rate—the rate of return on capital.

The return on capital after repayment of the loan is calculated as follows:

$$ROE = -PMT(Rate, Nper, E, Fv, Type)$$

PMT, Rate, Nper, E, Fv, and Type has the same meaning as above.

11.9 Financial Evaluation

11.9.1 Calculation Period of Financial Evaluation

This is a BOT project. The permitted operation period, construction period and financial evaluation calculation period is 30 years, 5 years and 35 years, respectively.

11.9.2 Financial Benchmark Yield

According to Pakistan's electricity policy, the financial benchmark yield is 10%.

11.9.3 Estimation of On-grid Price

According to the 15-year loan repayment requirements, the grace period and the loan repayment period is 5 years and 10 years, respectively. The return rate of capital is 17%. According to Pakistan's power policy, when dividend withholding tax is not included, the price of repayment period is 10.3749 cents/(kW·h), and the price of electricity after repayment is 4.8983 cents/(kW·h), and the average price is 8.4461 cents/(kW·h) during the operating period. When the dividend withholding tax is included, the electricity price after repayment is 5.1481 cents/(kW·h), and the average electricity price during the operation period is 8.6908 cents/(kW·h). See Table 11.4 for electricity price analysis of QK Hydropower Station and Table 11.5 for financial indicators. When the dividend withholding tax is included, the electricity price of the loan repayment period is 10.6138 cents/(kW·h). The price of electricity after repayment is 8.6908 cents/(kW·h) during the operating period of 5.1481 cents/(kW·h).

Table 11.4　　Electricity price analysis of QK Hydropower Station

Price constitution	Loan repayment period (1th-10th)	Loan repayment period (11th-30th)
Capacity price/[cents/(kW·h)]	3489.0257	1625.7549
(1) Fixed O&M fee (local currency)	101.4345	133.8135
(2) Fixed O&M fee (foreign currency)	85.1354	122.2901
(3) Property insurance	141.4375	141.4375
(4) Oversea investment premium	38.8488	15.9966
(5) Repayment of principal and interest payment	1949.9243	2.1875
(6) ROE	776.9758	812.1242
(7) ROEDC	313.4849	313.4849
(8) Dividend withholding tax	81.7846	84.4207
Energy price, EPP/[cents/(kW·h)]		
(1) Variable O&M fee (local currency)	0.1092	0.1440
(2) Variable O&M fee (foreign currency)	0.0278	0.0400
(3) Water resources utilization fee	0.1540	0.1540
Reference energy price, CPP+EPP/[cents/(kW·h)]		
(1) Excluding dividend withholding tax	10.3719	4.8983
(2) Including dividend withholding tax	10.6138	5.1481
levelized price	8.4461(Excluding dividend withholding tax)	
	8.6908(Including dividend withholding tax)	

11.9 Financial Evaluation

Table 11.5 Summary of financial evaluation indicators of QK Hydropower Station

Serial number	Item	Unit	Indicator	Note
1	Total investment	10^4 dollars	175505	
1.1	Construction investment	10^4 dollars	135780	
1.2	Bank loan interest	10^4 dollars	21672	
1.3	Overseas loan insurance premium	10^4 dollars	13089	
1.4	Management fee	10^4 dollars	1313	
1.5	Commitment fee	10^4 dollars	1934	
1.6	Oversea investment insurance premium	10^4 dollars	1317	
1.7	Working capital	10^4 dollars	400	
2	On-grid price	Cents/(kW·h)	8.4461	Excluding dividend
3	Total sales income	10^4 dollars	654411	
4	Total operation cost	10^4 dollars	145997	
5	Total repayment of principal and interest	10^4 dollars	186983	
6	Profit index			
6.1	Project FIRR	%	13.54%	
6.2	FIRR on capital	%	17.00%	
6.3	Pack back time, PBT	year	9.65	
7	Loan period	year	15	
8	Financing situation			
8.1	Capital fund	10^4 dollars	43776	
8.2	Debt fund	10^4 dollars	131629	

Fill in the total cost estimation table (Table 11.6) according to the fixed and variable operation cost estimation table (Table 11.7 - Table 11.9). Calculate the financial income, profit and tax of the project according to the estimated on-grid price and fill in the profit and profit distribution table (Table 11.10). Based on above tables, fill in the cash flow statement of financial plan (Table 11.11), cash flow statement of project investment (Table 11.12), capital cash flow statement (Table 11.13) and balance sheet (Table 11.14), and calculate the corresponding financial indicators.

At present, the on-grid price of Pakistan hydropower station is 8.5 - 9.0 cents/(kW·h). The average tariff during the operation is lower than the current on-grid price. Therefore, this power station has a competitive on-grid price.

Table 11.6 Estimation of fixed operation cost and variable operation of QK Hydropower Station during operation period

unit: 10^4 dollars

Serial number	Item	1	2	3	4	5	6	7	8	9	10	11	12	13	14	15
One	Fixed operation cost	33398	33418	33438	33459	33480	33503	33526	33550	33575	33600	15291	15319	15347	15376	15406
1	Fixed O&M fee (local currency)	896	912	929	946	964	981	999	1018	1036	1055	1075	1095	1115	1135	1156
2	Fixed O&M fee (foreign currency)	733	751	769	787	806	825	845	865	886	907	929	951	974	997	1021
3	Fixed asset premium	1358	1358	1358	1358	1358	1358	1358	1358	1358	1358	1358	1358	1358	1358	1358
4	Overseas investment premium	439	424	410	395	380	366	351	336	322	307	293	278	263	249	234
5	Repayment of loan principal	9505	10171	10883	11644	12459	13332	14265	15263	16332	17475	0	0	0	0	0
6	Repayment of loan interest	9214	8549	7837	7075	6260	5388	4454	3456	2387	1244	21	21	21	21	21
7	ROE (loan repayment period)	7459	7459	7459	7459	7459	7459	7459	7459	7459	7459	0	0	0	0	0
	ROE (after loan repayment)		0	0	0	0	0	0	0	0	0	7796	7796	7796	7796	7796
8	ROEDC	3009	3009	3009	3009	3009	3009	3009	3009	3009	3009	3009	3009	3009	3009	3009
9	Bonus withholding tax	785	785	785	785	785	785	785	785	785	785	810	810	810	810	810
Two	Variable operation cost	907	915	923	931	939	948	957	966	975	984	993	1003	1013	1023	1033
1	Variable O&M fee (local currency)	326	332	338	344	351	357	364	370	377	384	391	398	406	413	421
2	Variable O&M fee (foreign currency)	81	83	85	87	89	91	93	96	98	100	103	105	108	110	113
3	water resources fee	500	500	500	500	500	500	500	500	500	500	500	500	500	500	500

11.9 Financial Evaluation

Series number	Operation period	16	17	18	19	20	21	22	23	24	25	26	27	28	29	30
One	Fixed operation cost	15437	15469	15502	15536	15571	15608	15645	15683	15723	15763	15805	15848	15892	15938	15985
1	Fixed O&M fee (local currency)	1177	1199	1221	1243	1266	1289	1313	1337	1362	1387	1412	1438	1464	1491	1519
2	Fixed O&M fee (foreign currency)	1046	1071	1096	1123	1149	1177	1205	1234	1264	1294	1325	1357	1389	1422	1457
3	Fixed asset premium	1358	1358	1358	1358	1358	1358	1358	1358	1358	1358	1358	1358	1358	1358	1358
4	Overseas investment premium	219	205	190	176	161	146	132	117	102	88	73	59	44	29	15
5	Repayment of loan principal	0	0	0	0	0	0	0	0	0	0	0	0	0	0	0
6	Repayment of loan interest	21	21	21	21	21	21	21	21	21	21	21	21	21	21	21
7	ROE (loan repayment period)	0	0	0	0	0	0	0	0	0	0	0	0	0	0	0
	ROE (after loan repayment)	7796	7796	7796	7796	7796	7796	7796	7796	7796	7796	7796	7796	7796	7796	7796
8	ROEDC	3009	3009	3009	3009	3009	3009	3009	3009	3009	3009	3009	3009	3009	3009	3009
9	Bonus withholding tax	810	810	810	810	810	810	810	810	810	810	810	810	810	810	810
Two	Variable operation cost	1044	1054	1065	1076	1087	1099	1111	1123	1135	1147	1160	1173	1186	1199	1213
1	Variable O&M (local currency)	428	436	444	452	461	469	478	486	495	504	514	523	533	543	553
2	Variable O&M (foreign currency)	116	118	121	124	127	130	133	136	140	143	146	150	154	157	161
3	Water resources utilization fee	500	500	500	500	500	500	500	500	500	500	500	500	500	500	500

Note: Calculation starts from the year of operation period.

11 Foreign Case of Financial Evaluation of Water Conservancy and Hydropower Projects

Table 11.7 Calculation of on-grid price of QK Hydropower Station

Serial number of year	Energy price [cents/(kW·h)]				Capacity price [cents/(kW/month)]									Capacity price /[cents /(kW·h)]	Total energy price [cents/(kW·h)]	
	Variable operation cost (local currency)	Variable operation cost (foreign currency)	Water resources fee	Total	Fixed operation cost (local currency)	Fixed operation cost (foreign currency)	Property insurance	Overseas investment premium	Loan principal repayment	Pay of loan interest	ROE	Return on capital during construction period	Bonus withholding tax	Total		Including dividend tax	Excluding dividend tax
1	0.100	0.025	0.154	0.279	93.3	76.4	141.4	45.7	990.1	959.8	777.0	313.5	81.8	3,479.0	10.293	10.573	10.331
2	0.102	0.026	0.154	0.282	95.0	78.2	141.4	44.2	1,059.4	890.5	777.0	313.5	81.8	3,481.0	10.299	10.581	10.339
3	0.104	0.026	0.154	0.284	96.8	80.1	141.4	42.7	1,133.6	816.3	777.0	313.5	81.8	3,483.1	10.305	10.590	10.348
4	0.106	0.027	0.154	0.287	98.6	82.0	141.4	41.1	1,213.0	737.0	777.0	313.5	81.8	3,485.3	10.312	10.599	10.357
5	0.108	0.027	0.154	0.290	100.4	83.9	141.4	39.6	1,297.9	652.1	777.0	313.5	81.8	3,487.5	10.318	10.608	10.366
6	0.110	0.028	0.154	0.292	102.2	86.0	141.4	38.1	1,388.7	561.2	777.0	313.5	81.8	3,489.9	10.325	10.617	10.375
7	0.112	0.029	0.154	0.295	104.1	88.0	141.4	36.6	1,485.9	464.0	777.0	313.5	81.8	3,492.3	10.332	10.627	10.385
8	0.114	0.029	0.154	0.298	106.0	90.1	141.4	35.0	1,589.9	360.0	777.0	313.5	81.8	3,494.8	10.340	10.637	10.395
9	0.116	0.030	0.154	0.300	108.0	92.3	141.4	33.5	1,701.2	248.7	777.0	313.5	81.8	3,497.4	10.347	10.648	10.406
10	0.118	0.031	0.154	0.303	109.9	94.5	141.4	32.0	1,820.3	129.6	777.0	313.5	81.8	3,500.0	10.355	10.659	10.417
11	0.121	0.032	0.154	0.306	112.0	96.8	141.4	30.5	0.0	2.2	812.1	313.5	84.4	1,592.8	4.713	5.019	4.769
12	0.123	0.032	0.154	0.309	114.0	99.1	141.4	28.9	0.0	2.2	812.1	313.5	84.4	1,595.7	4.721	5.030	4.780
13	0.125	0.033	0.154	0.312	116.1	101.4	141.4	27.4	0.0	2.2	812.1	313.5	84.4	1,598.6	4.730	5.042	4.792
14	0.127	0.034	0.154	0.315	118.2	103.9	141.4	25.9	0.0	2.2	812.1	313.5	84.4	1,601.7	4.739	5.054	4.804
15	0.130	0.035	0.154	0.318	120.4	106.4	141.4	24.4	0.0	2.2	812.1	313.5	84.4	1,604.8	4.748	5.066	4.817
16	0.132	0.036	0.154	0.322	122.6	108.9	141.4	22.9	0.0	2.2	812.1	313.5	84.4	1,608.0	4.758	5.079	4.829
17	0.134	0.036	0.154	0.325	124.9	111.5	141.4	21.3	0.0	2.2	812.1	313.5	84.4	1,611.4	4.767	5.092	4.843
18	0.137	0.037	0.154	0.328	127.2	114.2	141.4	19.8	0.0	2.2	812.1	313.5	84.4	1,614.8	4.778	5.106	4.856
19	0.139	0.038	0.154	0.332	129.5	116.9	141.4	18.3	0.0	2.2	812.1	313.5	84.4	1,618.4	4.788	5.120	4.870

11.9 Financial Evaluation

Continued

Serial number of year	Energy price [cents/(kW·h)]					Capacity price [cents/(kW/month)]									Capacity price /[cents/(kW·h)]	Total energy price [cents/(kW·h)]	
	Variable operation cost (local currency)	Variable operation cost (foreign currency)	Water resources fee	Total		Fixed operation cost (local currency)	Fixed operation cost (foreign currency)	Property insurance	Overseas investment premium	Loan principal repayment	Pay of loan interest	ROE	Return on capital during construction period	Bonus withholding tax	Total		Including dividend tax	Excluding dividend tax
20	0.142	0.039	0.154	0.335		131.9	119.7	141.4	16.8	0.0	2.2	812.1	313.5	84.4	1,622.0	4.799	5.134	4.884
21	0.145	0.040	0.154	0.339		134.3	122.6	141.4	15.2	0.0	2.2	812.1	313.5	84.4	1,625.8	4.810	5.149	4.899
22	0.147	0.041	0.154	0.342		136.8	125.5	141.4	13.7	0.0	2.2	812.1	313.5	84.4	1,629.7	4.822	5.164	4.914
23	0.150	0.042	0.154	0.346		139.3	128.5	141.4	12.2	0.0	2.2	812.1	313.5	84.4	1,633.7	4.833	5.179	4.930
24	0.153	0.043	0.154	0.350		141.8	131.6	141.4	10.7	0.0	2.2	812.1	313.5	84.4	1,637.8	4.846	5.195	4.946
25	0.155	0.044	0.154	0.354		144.4	134.8	141.4	9.1	0.0	2.2	812.1	313.5	84.4	1,642.0	4.858	5.212	4.962
26	0.158	0.045	0.154	0.357		147.1	138.0	141.4	7.6	0.0	2.2	812.1	313.5	84.4	1,646.4	4.871	5.228	4.979
27	0.161	0.046	0.154	0.361		149.8	141.3	141.4	6.1	0.0	2.2	812.1	313.5	84.4	1,650.9	4.884	5.246	4.996
28	0.164	0.047	0.154	0.366		152.5	144.7	141.4	4.6	0.0	2.2	812.1	313.5	84.4	1,655.5	4.898	5.263	5.014
29	0.167	0.048	0.154	0.370		155.3	148.2	141.4	3.0	0.0	2.2	812.1	313.5	84.4	1,660.2	4.912	5.282	5.032
30	0.170	0.050	0.154	0.374		158.2	151.7	141.4	1.5	0.0	2.2	812.1	313.5	84.4	1,665.1	4.926	5.300	5.051
1–10 Years	0.109	0.028	0.154	0.291		101.4	85.1	141.4	38.8	1,368.0	581.9	777.0	313.5	81.8	3,489.0	10.323	10.6138	10.3719
11–30 Years	0.144	0.040	0.154	0.338		133.8	122.3	141.4	16.0	0.0	2.2	812.1	313.5	84.4	1,625.8	4.810	5.1481	4.8983
1–30 Years	0.132	0.036	0.154	0.322		123.0	109.9	141.4	23.6	456.0	195.4	800.4	313.5	83.5	2,246.8	6.648	6.9700	6.7228
Results after considering the social discount rate, 10%																		
1–30 Years	0.118	0.031	0.154	0.302		109.3	94.1	141.4	33.2	845.8	426.0	789.2	313.5	82.7	2,835.2	8.388	8.6908	8.4461

Note: Calculation starts from the year of operation period.

11 Foreign Case of Financial Evaluation of Water Conservancy and Hydropower Projects

Table 11.8 Loan repayment schedule of QK Hydropower Station

unit: 10^4 dollars

Serial number	Item	Total	1	2	3	4	5	6	7	8	9	10	11	12	13	14	15
1	Loan																
1.1	Loan balance at the beginning of the year		0	21907	51187	81945	108556	131329	121824	111653	100770	89126	76667	63335	49070	33807	17475
1.2	Repayment of principal and interest for the year	186983						18698	18698	18698	18698	18698	18698	18698	18698	18698	18698
	Repayment of principal	131329	0	0	0	0	0	9505	10171	10883	11644	12459	13332	14265	15263	16332	17475
	Payment of interest	55654	0	0	0	0	0	9193	8528	7816	7054	6239	5367	4433	3435	2366	1223
1.3	Loan balance at the end of the year		21907	51187	81945	108556	131329	121824	111653	100770	89126	76667	63335	49070	33807	17475	0
	Calculation indicators																
	Interest coverage ratio /%							2.54	2.74	2.99	3.31	3.74	4.34	5.26	6.77	9.81	18.81
	Debt service coverage ratio/%							1.76	1.80	1.83	1.87	1.92	1.96	2.01	2.07	2.13	2.19

Table 11.9 Total cost estimation of QK Hydropower Station

unit: 10^4 dollars

Serial number	Item	Total	1	2	3	4	5	6	7	8	9	10	11	12	13	14	15	16	17
1	Fixed O&M fee	67083	0	0	0	0	0	1629	1663	1698	1733	1769	1806	1844	1883	1922	1962	2004	2046
2	Fixed assets insurance premium	40734	0	0	0	0	0	1358	1358	1358	1358	1358	1358	1358	1358	1358	1358	1358	1358
3	Overseas investment insurance premium	6801	0	0	0	0	0	439	424	410	395	380	366	351	336	322	307	293	278
4	Variable O&M fee	16389	0	0	0	0	0	407	415	423	431	440	448	457	466	475	484	494	503
5	Water resources utilization fee	14991	0	0	0	0	0	500	500	500	500	500	500	500	500	500	500	500	500
6	Operation cost	145997	0	0	0	0	0	4332	4360	4388	4417	4447	4478	4510	4543	4576	4611	4647	4684

11.9 Financial Evaluation

Continued

Serial number	Item	Total	1	2	3	4	5	6	7	8	9	10	11	12	13	14	15	16	17
7	Depreciation cost	173354	0	0	0	0	0	5778	5778	5778	5778	5778	5778	5778	5778	5778	5778	5778	5778
8	Amortization fee	0	0	0	0	0	0	0	0	0	0	0	0	0	0	0	0	0	0
9	Interest expenditure	56284	0	0	0	0	0	9214	8549	7837	7075	6260	5388	4454	3456	2387	1244	0	0
9.1	Long-term loan interest expenditure	55654	0	0	0	0	0	9193	8528	7816	7054	6239	5367	4433	3435	2366	1223	0	0
9.2	Interest expenditure of working capital loan	630	0	0	0	0	0	21	21	21	21	21	21	21	21	21	21	21	21
10	Total cost	375635	0	0	0	0	0	19325	18687	18003	17270	16485	15644	14743	13777	12742	11634	10447	10484

Serial number	Item	18	19	20	21	22	23	24	25	26	27	28	29	30	31	32	33	34	35
1	Fixed O&M fee	2088	2132	2177	2223	2269	2317	2366	2415	2466	2518	2571	2625	2680	2737	2795	2853	2914	2975
2	Fixed asset insurance premium	1358	1358	1358	1358	1358	1358	1358	1358	1358	1358	1358	1358	1358	1358	1358	1358	1358	1358
3	Overseas investment insurance premium	263	249	234	219	205	190	176	161	146	132	117	102	88	73	59	44	29	15
4	Variable O&M fee	513	523	533	544	554	565	576	588	599	611	623	635	647	660	673	686	700	713
5	Water resources utilization fee	500	500	500	500	500	500	500	500	500	500	500	500	500	500	500	500	500	500
6	Operation cost	4722	4762	4802	4843	4886	4930	4975	5021	5069	5118	5168	5220	5273	5328	5384	5441	5500	5561
7	Depreciation cost	5778	5778	5778	5778	5778	5778	5778	5778	5778	5778	5778	5778	5778	5778	5778	5778	5778	5778
8	Amortization fee	0	0	0	0	0	0	0	0	0	0	0	0	0	0	0	0	0	0
9	Interest expenditure	21	21	21	21	21	21	21	21	21	21	21	21	21	21	21	21	21	21
9.1	Long-term loan interest expenditure	0	0	0	0	0	0	0	0	0	0	0	0	0	0	0	0	0	0
9.2	Interest expenditure of working capital loan	21	21	21	21	21	21	21	21	21	21	21	21	21	21	21	21	21	21
10	Total cost	10522	10561	10601	10643	10686	10729	10775	10821	10869	10918	10968	11020	11073	11127	11183	11241	11300	11360

Table 11.10　Profit and profit distribution statement

unit: 10^4 dollars

Serial number	Item	Total	1	2	3	4	5	6	7	8	9	10	11	12	13	14	15	16	17
1	Sales income	654411						33520	33547	33575	33605	33635	33666	33697	33730	33764	33799	15474	15511
1.1	Effective power/(GW·h)	97342						3245	3245	3245	3245	3245	3245	3245	3245	3245	3245	3245	3245
1.2	Effective capacity/MW							800	800	800	800	800	800	800	800	800	800	800	800
1.3	Energy price [cents/(kW·h)]							0.2794	0.2819	0.2844	0.2869	0.2895	0.2921	0.2948	0.2976	0.3004	0.3032	0.3062	0.3091
1.4	Capacity price [cents/(kW·h)]							3397	3399	3401	3404	3406	3408	3410	3413	3416	3418	1508	1511
2	Total cost	375635						19325	18687	18003	17270	16485	15644	14743	13777	12742	11634	10447	10484
3	Total profits	278776						14195	14861	15573	16334	17149	18022	18955	19953	21022	22165	5027	5027
4	Taxable amount of income	0																	
5	Income tax	0																	
6	Net profit (5−8)	278776						14195	14861	15573	16334	17149	18022	18955	19953	21022	22165	5027	5027
7	Undistributed profit at the beginning of the period	0																	
8	Profit available for distribution	278776						14195	14861	15573	16334	17149	18022	18955	19953	21022	22165	5027	5027
9	Withdrawal of statutory surplus reserve	0																	
10	Profit available for distribution to investors	278776						14195	14861	15573	16334	17149	18022	18955	19953	21022	22165	5027	5027
11	Profit distribution by investors	320801						10468	10468	10468	10468	10468	10468	10468	10468	10468	10468	10806	10806
12	Undistributed profit	−42025						3727	4392	5104	5866	6681	7553	8486	9485	10553	11697	−5778	−5778
13	Profit before interest and tax	335060						23409	23409	23409	23409	23409	23409	23409	23409	23409	23409	5048	5048
14	Profit before interest, tax, depreciation and amortization	466389						32914	35580	34292	35054	35869	36741	37674	38673	39741	40884	5048	5048

11.9 Financial Evaluation

Continued

Serial number	Item	18	19	20	21	22	23	24	25	26	27	28	29	30	31	32	33	34	35
1	Sales income	15549	15588	15629	15670	15713	15757	15802	15848	15896	15945	15995	16047	16100	16155	16211	16268	16327	16388
1.1	Effective power/(GW·h)	3245	3245	3245	3245	3245	3245	3245	3245	3245	3245	3245	3245	3245	3245	3245	3245	3245	3245
1.2	Effective capacity/MW	800	800	800	800	800	800	800	800	800	800	800	800	800	800	800	800	800	800
1.3	Energy price/[cents/(kW·h)]	0.3122	0.3152	0.3184	0.3216	0.3249	0.3282	0.3316	0.3351	0.3387	0.3423	0.3460	0.3497	0.3535	0.3575	0.3614	0.3655	0.3697	0.3739
1.4	Capacity price/[cents/(kW·h)]	1514	1517	1520	1524	1527	1530	1534	1538	1541	1545	1549	1553	1558	1562	1566	1571	1576	1581
2	Total cost	10522	10561	10601	10643	10686	10729	10775	10821	10869	10918	10968	11020	11073	11127	11183	11241	11300	11360
3	Total profits	5027	5027	5027	5027	5027	5027	5027	5027	5027	5027	5027	5027	5027	5027	5027	5027	5027	5027
4	Taxable income																		
5	Income tax																		
6	Net profit (5−8)	5027	5027	5027	5027	5027	5027	5027	5027	5027	5027	5027	5027	5027	5027	5027	5027	5027	5027
7	Undistributed profit at the beginning of the period																		
8	Profit available for distribution	5027	5027	5027	5027	5027	5027	5027	5027	5027	5027	5027	5027	5027	5027	5027	5027	5027	5027
9	Withdrawal of statutory surplus reserve																		
10	Profit available for distribution to investors	5027	5027	5027	5027	5027	5027	5027	5027	5027	5027	5027	5027	5027	5027	5027	5027	5027	5027
11	Profit distribution by investors	10806	10806	10806	10806	10806	10806	10806	10806	10806	10806	10806	10806	10806	10806	10806	10806	10806	10806
12	Undistributed profit	−5778	−5778	−5778	−5778	−5778	−5778	−5778	−5778	−5778	−5778	−5778	−5778	−5778	−5778	−5778	−5778	−5778	−5778
13	Profit before interest and tax	5048	5048	5048	5048	5048	5048	5048	5048	5048	5048	5048	5048	5048	5048	5048	5048	5048	5048
14	Profit before interest, tax, depreciation and amortization	5048	5048	5048	5048	5048	5048	5048	5048	5048	5048	5048	5048	5048	5048	5048	5048	5048	5048

Note: After loan repayment, the undistributed profit is negative.

11 Foreign Case of Financial Evaluation of Water Conservancy and Hydropower Projects

Table 11.11 Cash flow statement of financial plan of QK Hydropower Station

unit: 10⁴ dollars

Serial number	Item	Total totaltotal	1	2	3	4	5	6	7	8	9	10	11	12	13	14	15	16	17
1	Net cash of operation	508414	0	0	0	0	0	29188	29188	29188	29188	29188	29188	29188	29188	29188	29188	10827	10827
1.1	Cash inflow	654411	0	0	0	0	0	33520	33547	33575	33605	33635	33666	33697	33730	33764	33799	15474	15511
1.1.1	Sales income	654411	0	0	0	0	0	33520	33547	33575	33605	33635	33666	33697	33730	33764	33799	15474	15511
1.1.2	Other income	0	0	0	0	0	0	0	0	0	0	0	0	0	0	0	0	0	0
1.2	Cash outflow	145997	0	0	0	0	0	4332	4360	4388	4417	4447	4478	4510	4543	4576	4611	4647	4684
1.2.1	Operation cost	145997	0	0	0	0	0	4332	4360	4388	4417	4447	4478	4510	4543	4576	4611	4647	4684
1.2.2	Income tax	0	0	0	0	0	0	0	0	0	0	0	0	0	0	0	0	0	0
1.2.3	Other outflows	0	0	0	0	0	0	0	0	0	0	0	0	0	0	0	0	0	0
2	Net cash flow from investment activities	−136180	−13655	−35828	−35917	−28552	−22228	0	0	0	0	0	0	0	0	0	0	0	0
2.1	Cash inflow	0																	
2.2	Cash outflow	136180	13655	35828	35917	28552	22228	0	0	0	0	0	0	0	0	0	0	0	0
2.2.1	Construction investment	135780	13655	35828	35917	28552	21828	0	0	0	0	0	0	0	0	0	0	0	0
2.2.2	Maintain operation investment	0	0	0	0	0	0	0	0	0	0	0	0	0	0	0	0	0	0
2.2.3	Working capital	400	0	0	0	0	400	0	0	0	0	0	0	0	0	0	0	0	0
2.2.4	Other outflows	0	0	0	0	0	0	0	0	0	0	0	0	0	0	0	0	0	0
3	Net cash flow from financing	−372234	13655	35828	35917	28552	22228	−29188	−29188	−29188	−29188	−29188	−29188	−29188	−29188	−29188	−29188	−10827	−10827

11.9 Financial Evaluation

Continued

Serial number	Item	Total totaltotal	1	2	3	4	5	6	7	8	9	10	11	12	13	14	15	16	17
3.1	Net cash inflow	175505	29209	39040	41011	35481	30764	0	0	0	0	0	0	0	0	0	0	0	0
3.1.1	Project capital investment	43876	7302	9760	10253	8870	7691	0	0	0	0	0	0	0	0	0	0	0	0
3.1.2	Project investment loan	131329	21907	29280	30758	26611	22773	0	0	0	0	0	0	0	0	0	0	0	0
3.1.3	Short term loan	0	0	0	0	0	0	0	0	0	0	0	0	0	0	0	0	0	0
3.1.4	Bonds	0	0	0	0	0	0	0	0	0	0	0	0	0	0	0	0	0	0
3.1.5	Working capital loan	300	0	0	0	0	300	0	0	0	0	0	0	0	0	0	0	0	0
3.1.6	Other income	0	0	0	0	0	0	0	0	0	0	0	0	0	0	0	0	0	0
3.2	cash outflow	547739	15554	3212	5094	6929	8536	29188	29188	29188	29188	29188	29188	29188	29188	29188	29188	10827	10827
3.2.1	Various interest expenses	95609	15554	3212	5094	6929	8536	9214	8549	7837	7075	6260	5388	4454	3456	2387	1244	21	21
3.2.2	Repayment of debt principal	131329	0	0	0	0	0	9505	10171	10883	11644	12459	13332	14265	15263	16332	17475	0	0
3.2.3	Profit payable (dividend distribution)	320801	0	0	0	0	0	10468	10468	10468	10468	10468	10468	10468	10468	10468	10468	10806	10806
3.2.4	Other outflows	0	0	0	0	0	0	0	0	0	0	0	0	0	0	0	0	0	0
4	Net cash flow	0	0	0	0	0	0	0	0	0	0	0	0	0	0	0	0	0	0
5	Accumulated surplus capital																		

Continued

Serial number	Item	18	19	20	21	22	23	24	25	26	27	28	29	30	31	32	33	34	35
1	Net cash of operation	10827	10827	10827	10827	10827	10827	10827	10827	10827	10827	10827	10827	10827	10827	10827	10827	10827	10827
1.1	Cash inflow	15549	15588	15629	15670	15713	15757	15802	15848	15896	15945	15995	16047	16100	16155	16211	16268	16327	16388
1.1.1	Sales income	15549	15588	15629	15670	15713	15757	15802	15848	15896	15945	15995	16047	16100	16155	16211	16268	16327	16388
1.1.2	Other income	0	0	0	0	0	0	0	0	0	0	0	0	0	0	0	0	0	0
1.2	Cash outflows	4722	4762	4802	4843	4886	4930	4975	5021	5069	5118	5168	5220	5273	5328	5384	5441	5500	5561
1.2.1	Operation cost	4722	4762	4802	4843	4886	4930	4975	5021	5069	5118	5168	5220	5273	5328	5384	5441	5500	5561
1.2.2	Income tax	0	0	0	0	0	0	0	0	0	0	0	0	0	0	0	0	0	0
1.2.3	Other outflows	0	0	0	0	0	0	0	0	0	0	0	0	0	0	0	0	0	0
2	Net cash flow from investment activities	0	0	0	0	0	0	0	0	0	0	0	0	0	0	0	0	0	0
2.1	Cash inflow																		
2.2	Cash outflow	0	0	0	0	0	0	0	0	0	0	0	0	0	0	0	0	0	0
2.2.1	Construction investment	0	0	0	0	0	0	0	0	0	0	0	0	0	0	0	0	0	0
2.2.2	Maintain operation investment	0	0	0	0	0	0	0	0	0	0	0	0	0	0	0	0	0	0
2.2.3	Working capital	0	0	0	0	0	0	0	0	0	0	0	0	0	0	0	0	0	0
2.2.4	Other outflows	0	0	0	0	0	0	0	0	0	0	0	0	0	0	0	0	0	0
3	Net cash flow from financing	−10827	−10827	−10827	−10827	−10827	−10827	−10827	−10827	−10827	−10827	−10827	−10827	−10827	−10827	−10827	−10827	−10827	−10827

11.9 Financial Evaluation

Continued

Serial number	Item	18	19	20	21	22	23	24	25	26	27	28	29	30	31	32	33	34	35
3.1	Cash inflow	0	0	0	0	0	0	0	0	0	0	0	0	0	0	0	0	0	0
3.1.1	Project capital investment	0	0	0	0	0	0	0	0	0	0	0	0	0	0	0	0	0	0
3.1.2	Project investment loan	0	0	0	0	0	0	0	0	0	0	0	0	0	0	0	0	0	0
3.1.3	Short term loan	0	0	0	0	0	0	0	0	0	0	0	0	0	0	0	0	0	0
3.1.4	Bonds	0	0	0	0	0	0	0	0	0	0	0	0	0	0	0	0	0	0
3.1.5	Working capital loan	0	0	0	0	0	0	0	0	0	0	0	0	0	0	0	0	0	0
3.1.6	Other inflow	0	0	0	0	0	0	0	0	0	0	0	0	0	0	0	0	0	0
3.2	Cash outflow	10827	10827	10827	10827	10827	10827	10827	10827	10827	10827	10827	10827	10827	10827	10827	10827	10827	10827
3.2.1	Various interest expenses	21	21	21	21	21	21	21	21	21	21	21	21	21	21	21	21	21	21
3.2.2	Repayment of debt principal	0	0	0	0	0	0	0	0	0	0	0	0	0	0	0	0	0	0
3.2.3	Profit payable (dividend distribution)	10806	10806	10806	10806	10806	10806	10806	10806	10806	10806	10806	10806	10806	10806	10806	10806	10806	10806
3.2.4	Other outflows	0	0	0	0	0	0	0	0	0	0	0	0	0	0	0	0	0	0
4	Net cash flow	0	0	0	0	0	0	0	0	0	0	0	0	0	0	0	0	0	0
5	Accumulated surplus capital	0	0	0	0	0	0	0	0	0	0	0	0	0	0	0	0	0	0

11 Foreign Case of Financial Evaluation of Water Conservancy and Hydropower Projects

Table 11.12 Project investment cash flow statement of QK Hydropower Station

unit: 10⁴ dollars

Serial number	Item	Total	1	2	3	4	5	6	7	8	9	10	11	12	13	14	15	16	17
1	Cash inflow	654411	0	0	0	0	0	33520	33547	33575	33605	33635	33666	33697	33730	33764	33799	15474	15511
1.1	Sales income	654411						33520	33547	33575	33605	33635	33666	33697	33730	33764	33799	15474	15511
1.2	Recovery of residual value of fixed assets	0	0	0	0	0	0	0	0	0	0	0	0	0	0	0	0	0	0
1.3	Recovery of working capital	0	0	0	0	0	0	0	0	0	0	0	0	0	0	0	0	0	0
2	Cash outflow	281777	13655	35828	35917	28552	21828	4332	4360	4388	4417	4447	4478	4510	4543	4576	4611	4647	4684
2.1	Construction investment	135780	13655	35828	35917	28552	21828												
2.2	Operation cost	145997						4332	4360	4388	4417	4447	4478	4510	4543	4576	4611	4647	4684
2.3	Dividend tax	0																	
2.4	Income tax	0																	
2.5	Maintain operational investment	0																	
3	Net cash flow	372634	−13655	−35828	−35917	−28552	−21828	29188	29188	29188	29188	29188	29188	29188	29188	29188	29188	10827	10827
4	Accumulated net cash flow		−13655	−49483	−85400	−113952	−135780	−106592	−77405	−48217	−19029	10158	39346	68534	97722	126909	156097	166924	177751

11.9 Financial Evaluation

Continued

Serial number	Item	18	19	20	21	22	23	24	25	26	27	28	29	30	31	32	33	34	35
1	Cash inflow	15549	15588	15629	15670	15713	15757	15802	15848	15896	15945	15995	16047	16100	16155	16211	16268	16327	16388
1.1	Sales income	15549	15588	15629	15670	15713	15757	15802	15848	15896	15945	15995	16047	16100	16155	16211	16268	16327	16388
1.2	Recovery of residual value of fixed assets	0	0	0	0	0	0	0	0	0	0	0	0	0	0	0	0	0	0
1.3	Recovery of working capital	0	0	0	0	0	0	0	0	0	0	0	0	0	0	0	0	0	0
2	Cash outflow	4722	4762	4802	4843	4886	4930	4975	5021	5069	5118	5168	5220	5273	5328	5384	5441	5500	5561
2.1	Construction investment																		
2.2	Operation cost	4722	4762	4802	4843	4886	4930	4975	5021	5069	5118	5168	5220	5273	5328	5384	5441	5500	5561
2.3	Dividend tax																		
2.4	Income tax																		
2.5	Maintain operation investment																		
3	Net cash flow (1 − 2)	10827	10827	10827	10827	10827	10827	10827	10827	10827	10827	10827	10827	10827	10827	10827	10827	10827	10827
4	Accumulated net cash flow	188578	199404	210231	221058	231885	242712	253539	264365	275192	286019	296846	307673	318500	329327	340153	350980	361807	372634

Evaluation indicators:

FIRR of project investment: 13.54%

FNPV of project investment ($i_c = 10\%$): 313.62 million yuan

Payback period of project investment (year): 9.65

11 Foreign Case of Financial Evaluation of Water Conservancy and Hydropower Projects

Table 11.13 Capital cash flow statement of QK Hydropower Station

unit: 10^4 dollars

Serial number	Item	Total	1	2	3	4	5	6	7	8	9	10	11	12	13	14	15	16	17
1	Cash inflow	654411	0	0	0	0	0	33520	33547	33575	33605	33635	33666	33697	33730	33764	33799	15474	15511
1.1	Sales income	654411						33520	33547	33575	33605	33635	33666	33697	33730	33764	33799	15474	15511
1.2	Recovery of residual value of fixed assets	0	0	0	0	0	0	0	0	0	0	0	0	0	0	0	0	0	0
1.3	Recovery of working capital	0	0	0	0	0	0	0	0	0	0	0	0	0	0	0	0	0	0
2	Cash outflow	370184	7302	9760	10253	8870	7691	23052	23079	23107	23136	23166	23197	23229	23262	23296	23331	4668	4705
2.1	Project capital	36574	7302	9760	10253	8870	7691												
2.2	Repayment of loan principal	131329						9505	10171	10883	11644	12459	13332	14265	15263	16332	17475	0	0
2.3	Loan interest payment	56284						9214	8549	7837	7075	6260	5388	4454	3456	2387	1244	21	21
2.4	Operation cost	145997						4332	4360	4388	4417	4447	4478	4510	4543	4576	4611	4647	4684
2.5	Dividend tax	0																	
2.6	Income tax	0																	
2.7	Maintain operational investment																		
3	Net cash flow	284227	−7302	−9760	−10253	−8870	−7691	10468	10468	10468	10468	10468	10468	10468	10468	10468	10468	10806	10806
4	Accumulated net cash flow		−7302	−17062	−27315	−36185	−43876	−33408	−22939	−12471	−2003	8466	18934	29403	39871	50340	60808	71614	82420

11.9 Financial Evaluation

Continued

Serial number	Item	18	19	20	21	22	23	24	25	26	27	28	29	30	31	32	33	34	35
1	Cash inflow	15549	15588	15629	15670	15713	15757	15802	15848	15896	15945	15995	16047	16100	16155	16211	16268	16327	16388
1.1	Sales income	15549	15588	15629	15670	15713	15757	15802	15848	15896	15945	15995	16047	16100	16155	16211	16268	16327	16388
1.2	Recovery of residual value of fixed assets	0	0	0	0	0	0	0	0	0	0	0	0	0	0	0	0	0	0
1.3	Recovery of working capital	0	0	0	0	0	0	0	0	0	0	0	0	0	0	0	0	0	0
2	Cash outflow	4743	4783	4823	4864	4907	4951	4996	5042	5090	5139	5189	5241	5294	5349	5405	5462	5521	5582
2.1	Project capital																		
2.2	Repayment of loan principal	0	0	0	0	0	0	0	0	0	0	0	0	0	0	0	0	0	0
2.3	Loan interest payment	21	21	21	21	21	21	21	21	21	21	21	21	21	21	21	21	21	21
2.4	Operation cost	4722	4762	4802	4843	4886	4930	4975	5021	5069	5118	5168	5220	5273	5328	5384	5441	5500	5561
2.5	Dividend tax																		
2.6	Income tax																		
2.7	Maintain operational investment																		
3	Net cash flow	10806	10806	10806	10806	10806	10806	10806	10806	10806	10806	10806	10806	10806	10806	10806	10806	10806	10806
4	Accumulated net cash flow	93225	104031	114837	125643	136449	147255	158061	168866	179672	190478	201284	212090	222896	233701	244507	255313	266119	276925

Evaluation indicators:
FIRR on capital: 17.00%

Table 11.14 Balance sheet of QK Hydropower Station

unit: 10^4 dollars

Serial number	Item	Total	1	2	3	4	5	6	7	8	9	10	11	12	13	14	15	16	17
1	Assets	3105130	29209	68249	109260	144741	175505	169727	163948	158170	152391	146613	140834	135056	129277	123499	117720	111942	106163
1.1	Total amount of working capital	12400	0	0	0	0	400	400	400	400	400	400	400	400	400	400	400	400	400
1.1.1	Working capital	12400	0	0	0	0	400	400	400	400	400	400	400	400	400	400	400	400	400
1.1.2	Accumulated surplus capital	0																	
1.2	Project under construction	526564	29209	68249	109260	144741	175105	0	0	0	0	0	0	0	0	0	0	0	0
1.3	Net value of fixed assets	2566166						169327	163548	157770	151991	146213	140434	134656	128877	123099	117320	111542	105763
1.4	Net value of intangible assets and other assets	0																	
2	Liabilities and owner's equity	3105130	29209	68249	109260	144741	175505	169727	163948	158170	152391	146613	140834	135056	129277	123499	117720	111942	106163
2.1	Total current liabilities	0	0	0	0	0	0	0	0	0	0	0	0	0	0	0	0	0	0
2.2	Construction investment loan	1058650	21907	51187	81945	108556	131329	121824	111653	100770	89126	76667	63335	49070	33807	17475	0	0	0
2.3	Working capital loan	9300	0	0	0	0	300	300	300	300	300	300	300	300	300	300	300	300	300
2.4	Subtotal of loan	1067950	21907	51187	81945	108556	131629	122124	111953	101070	89426	76967	63635	49370	34107	17775	300	300	300
2.5	Owner's equity	2037180	7302	17062	27315	36185	43876	47603	51995	57099	62965	69646	77199	85686	95171	105724	117420	111642	105863
2.5.1	Capital fund	1448030	7302	17062	27315	36185	43876	43876	43876	43876	43876	43876	43876	43876	43876	43876	43876	43876	43876
2.5.2	Capital reserve	0																	
2.5.3	Accumulated surplus accumulation fund		0	0	0	0	0	0	0	0	0	0	0	0	0	0	0	0	0
2.5.4	Accumulated undistributed profits		0	0	0	0	0	3727	8119	13223	19089	25770	33323	41809	51294	61848	73544	67766	61987
	Asset liability ratio/%		75.0	75.00	75.00	75.00	75.00	71.95	68.29	63.90	58.68	52.50	45.18	36.56	26.38	14.39	0.25	0.27	0.28

11.9 Financial Evaluation

Continued

Serial number	Item	18	19	20	21	22	23	24	25	26	27	28	29	30	31	32	33	34	35
1	Capital	100385	94607	88828	83050	77271	71493	65714	59936	54157	48379	42600	36822	31043	25265	19486	13708	7930	2151
1.1	Total amount of working capital	400	400	400	400	400	400	400	400	400	400	400	400	400	400	400	400	400	400
1.1.1	Working capital	400	400	400	400	400	400	400	400	400	400	400	400	400	400	400	400	400	400
1.1.2	Accumulated surplus capital																		
1.2	Project under construction	0	0	0	0	0	0	0	0	0	0	0	0	0	0	0	0	0	0
1.3	Net value of fixed assets	99985	94207	88428	82650	76871	71093	65314	59536	53757	47979	42200	36422	30643	24865	19086	13308	7530	1751
1.4	Net value of intangible assets and other assets																		
2	Liabilities and owner's equity	100385	94607	88828	83050	77271	71493	65714	59936	54157	48379	42600	36822	31043	25265	19486	13708	7930	2151
2.1	Total current liabilities	0	0	0	0	0	0	0	0	0	0	0	0	0	0	0	0	0	0
2.2	Construction investment loan	0	0	0	0	0	0	0	0	0	0	0	0	0	0	0	0	0	0
2.3	Working capital loan	300	300	300	300	300	300	300	300	300	300	300	300	300	300	300	300	300	300
2.4	Subtotal of loan	300	300	300	300	300	300	300	300	300	300	300	300	300	300	300	300	300	300
2.5	Owner's equity	100085	94307	88528	82750	76971	71193	65414	59636	53857	48079	42300	36522	30743	24965	19186	13408	7630	1851
2.5.1	Capital fund	43876	43876	43876	43876	43876	43876	43876	43876	43876	43876	43876	43876	43876	43876	43876	43876	43876	43876
2.5.2	Capital reserve																		
2.5.3	Accumulated surplus accumulation fund	0	0	0	0	0	0	0	0	0	0	0	0	0	0	0	0	0	0
2.5.4	Accumulated undistributed profits	56209	50430	44652	38873	33095	27316	21538	15759	9981	4203	−1576	−7354	−13133	−18911	−24690	−30468	−36247	−42025
	Asset liability ratio /%	0.30	0.32	0.34	0.36	0.39	0.42	0.46	0.50	0.55	0.62	0.70	0.81	0.97	1.19	1.54	2.19	3.78	13.95

11.9.4 Financial Evaluation

1. Financial viability analysis

From the cash flow statement of financial plan, the project has enough net cash flow of operating activities to meet the requirements every year during the operation period and ensure the operation of the power station. This means the financial status of the project is sustainable and the financial viability is strong.

2. Solvency analysis

The capital of the project is 25% of the total investment, and the rest 75% of the total investment uses bank loans with an annual interest rate of 7%. The loan repayment period is 15 years (calculated from the first loan), including a grace period of 5 years and a repayment period of 10 years. The repayment method is to repay the principal and interest of the same amount once a year. See Table 11.8 for the calculation of repayment of principal and interest. It can be seen from the calculation table of loan repayment of principal and interest that: according to the recommended feed in tariff, the interest reserve ratio of the power station is 2.54 - 18.81 (greater than 2), and the debt service reserve ratio of the power station is 1.76 - 2.19 (greater than 1.3), indicating that the project has strong debt service ability.

It can be seen from the balance sheet that the highest asset liability ratio of QK Hydropower Station in the construction period is 75%. As the project is put into operation, the asset liability ratio gradually decreases. It would be less than 1% at the end of the loan repayment period. With the gradual decrease of assets, the asset liability ratio increases with the constant working capital, but the total liability is not large.

3. Profitability analysis

The power station keeps separate accounting. The power generation income is calculated as the on-grid electricity amount multiplied by on-grid electricity price. According to the profit and profit distribution statement, project investment cash flow statement and project capital cash flow statement, it can be seen that the electricity price during the loan repayment period is 10.3719 cents/(kW·h). The electricity price after the loan repayment is 4.8983 cents/(kW·h). The average electricity price during the operation period is 8.4461 cents/(kW·h). The total sales revenue of power generation during the whole calculation period is 6544.11 million dollars. According to the recommended feed in tariff, the FIRR of project investment and capital fund is 13.54% and 17.00%, respectively. The payback period of investment is 9.65 years. In summary, the project has good profitability.

11.10 Sensitivity Analysis

The sensitivity factor analysis considers the impacts of construction investment, on-grid energy, interest rate on the average electricity price during the operating period. See Table 11.15 for sensitivity analysis results.

Table 11.15　　　Sensitivity Analysis of Financial Evaluation

Serial number	Item	FIRR/%		Average electricity price during operation period /[cents/(kW·h)]
		All investment	Capital fund	
1	Basic plan	13.54	17.00	8.4461
2	Changes in construction investment			
2.1	Decrease by 20%	13.54	17.00	6.9401
2.2	Decrease by 10%	13.54	17.00	7.6931
2.3	Increase by 10%	13.54	17.00	9.1991
2.4	Increase by 20%	13.54	17.00	9.9521
3	Variation of on-grid energy			
3.1	Decrease by 20%	13.54	17.00	10.5191
3.2	Decrease by 10%	13.54	17.00	9.3674
3.3	Increase by 10%	13.54	17.00	7.6923
3.4	Increase by 20%	13.54	17.00	7.0641
4	Long-term loan interest rate			
4.1	5.0%	12.11	17.00	7.7836
4.2	5.5%	12.35	17.00	7.9423
4.3	6.0%	12.81	17.00	8.1055
4.4	6.5%	13.16	17.00	8.2734
4.5	7.5%	13.90	17.00	8.6237
4.6	8.0%	14.28	17.00	8.8065
5	The loan repayment period is 18 years	13.25	17.00	8.4198

It can be seen from the above table that when the construction investment increases or decreases by 20%, the change range of the on-grid price is 9.9521 - 6.9401 cents/(kW·h). When the on-grid energy increases or decreases by 20%, the on-grid price varies from 7.0641 - 10.5191 cents/(kW·h). When the loan interest rate is 5.0% - 8.0%, the on-grid price varies from 7.7836 - 8.8065 cents/(kW·h). When the loan repayment period is 18 years, the on-grid price is 8.4198 cents/(kW·h).

11.11 Conclusions

The construction investment of the project is 1357.8 million dollars (excluding the power transmission and transformation project). After considering the financing cost, the total investment is 1755.17 million dollars. When the return on capital is 17%, within the 30-year operation period, the average on-grid price of this project is 8.4461 cents/(kW·h) (excluding dividend withholding tax), according to the Pakistan Power Generation Project Policy (Year 2002) as well as the requirement of 15-year loan repayment. Compared to similar hydropower stations in Pakistan, the on-grid price of this power station has strong competitiveness.

According to the recommended on-grid price, the project can meet the operating requirements during the operation period every year. It means the project has a strong financial viability. The interest ratio of the power station is 2.54 - 18.81 (larger than 2). The debt service reserve ratio is 1.76 - 2.19 (larger than 1.3), implying that the debt service ability of the project is strong. The financial internal rate of return of the project investment and project capital is 13.54% and 17.00%, respectively. The investment recovery period is 9.65 years. This means the project has good profitability.

References

[1] National Development and Reform Commission, Ministry of Construction. Construction Projects Economic Evaluation Method and Parameter (3rd Edition) [M]. Beijing: China Planning Press, 2006.

[2] Compilation Group of *Study Guide of Investment Project Feasibility*. Study Guide of Investment Project Feasibility [M]. Beijing: China Electric Power Press, 2005.

[3] Regulation for Economic Evaluation of Water Conservancy Construction Projects (SL 72 - 2013) [S]. Beijing: China Water Power Press, 2014.

[4] Specification on Economic Evaluation of Hydropower Project (DL/T 5441 - 2010) [S]. Beijing: China Electric Power Press, 2010.

[5] Ministry of Electric Power Industry of The People's Republic of China. General Institute of Water Conservancy and Hydropower Planning and Design, Ministry of Water Resources of the People's Republic of China. Interim Provisions on financial evaluation of hydropower construction projects (Trial) [M]. Beijing: China Electric Power Press, 1994.

[6] Zhanpang Zhang, et al. Water economics [M]. Beijing: China Central Radio and Television University, 2002.

[7] Lisheng Suo, Ning Liu. Handbook of Hydraulic Structure Design [M]. Beijing: China Water Power Press, 2014.

[8] Shoufa Yu, etc. Practical book on economic evaluation methods and parameters of construction projects [M]. Beijing: China Building Materials Press, 1999.

[9] Liping Wang, et al. Water Conservancy Engineering Economics [M]. Wuhan: Wuhan University Press, 2002.

[10] National Survey and Design Registered Engineer Water Conservancy and Hydropower Engineering Management Committee. China Water Conservancy and Hydropower Investigation and Design Association. Professional cases of Water Conservancy and Hydropower Engineering (engineering planning, soil and water conservation and Engineering Resettlement) [M]. Zhengzhou: The Yellow River Water Conservancy Press, 2007.

[11] Reference Materials Compilation Committee of National Qualification Examination for Registered Consulting Engineer (investment). Project decision analysis and evaluation (2008 Edition) [M]. Beijing: China Planning Press, 2007.

[12] Guidelines for Calculating Ecological Benefit on the Project of Small Hydropower Substituting for Fuel (SL 593 - 2013) (SL 593 - 2013) [S]. Beijing: China Water Power Press, 2013.

[13] Comprehensive Control of Soil and Water Conservation - Method of Benefit Calculation (GB/T 15774 -2008) General Administration of Quality Supervision, Inspection and Quarantine of the People's Republic of China, National Standardization Administration of China. 2008 - 11 - 14.